Culinary Math
PRINCIPLES AND APPLICATIONS

AMERICAN TECHNICAL PUBLISHERS, INC.
ORLAND PARK, ILLINOIS 60467-5756

Chef Michael J. McGreal

Linda J. Padilla

Culinary Math Principles and Applications contains procedures commonly practiced in the foodservice industry. Specific procedures vary with each task and must be performed by a qualified person. For maximum safety, always refer to specific manufacturer recommendations, insurance regulations, specific facility procedures, applicable federal, state, and local regulations, and any authority having jurisdiction. The material contained is intended to be an educational resource for the user. American Technical Publishers, Inc. assumes no responsibility or liability in connection with this material or its use by any individual or organization.

American Technical Publishers, Inc., Editorial Staff

Editor in Chief:
        Jonathan F. Gosse
Vice President—Production:
        Peter A. Zurlis
Director of Product Development:
        Cathy A. Scruggs
Art Manager:
        James M. Clarke
Technical Editor:
        Michael A. Sodaro
Copy Editor:
        Talia J. Turner
Cover Design:
        James M. Clarke
Illustration/Layout:
        Jennifer M. Hines
Multimedia Coordinator:
        Carl R. Hansen
CD-ROM Development:
        Gretje Dahl
        Daniel Kundrat
        Nicole S. Polak
        Robert E. Stickley

1 2 3 4 5 6 7 8 9 – 10 – 9 8 7 6 5 4 3 2

Printed in the United States of America

        ISBN 978-0-8269-4211-1

 This book is printed on 30% recycled paper.

Acknowledgments

About the Authors

Michael J. McGreal, CEC, CCE, CHE, FMP, CHA, MCFE, is the Culinary Arts/Hospitality Management Department Chair at Joliet Junior College and the 2009 FENI Postsecondary Educator of the Year. Chef McGreal has over 25 years of foodservice experience and his first textbook, *Culinary Arts Principles and Applications,* received a 2008 Cordon d'Or International Culinary Academy Award.

Linda J. Padilla has taught mathematics at Joliet Junior College for over 25 years. She holds degrees in mathematics, education, and counseling. Linda has extensive experience collaborating with other disciplines and has developed curriculum strategies and applications for math in the culinary arts. Her work on math applications in the culinary field has also been featured at national conferences.

The authors and publisher are grateful for the technical reviews provided by the following individuals:

Tom Hickey, CEC, CCE, CFE, CHE, CCP
Director, National Center for Hospitality Studies
Sullivan University, Louisville, KY

Anthony Lowman, CCC, CCE
Culinary Apprenticeship Coordinator
First Coast Technical Institute, St. Augustine, FL

Christopher Plemmons, CEC, AAC
Chef Instructor
Olympic College, Bremerton, WA

The authors and publisher are grateful for the technical assistance provided by the following companies and organizations:

- The Beef Checkoff
- Browne-Halco (NJ)
- Carlisle FoodService Products
- Charlie Trotter's
- Classic Party Rentals
- Coopers-Atkins Corporation
- Cres Cor
- Daniel, NYC
- Detecto, A Division of Cardinal Scale Manufacturing Co.
- Edlund Co.
- Florida Tomato Committee
- Fluke Corporation
- Harbor Seafood
- Idaho Potato Commission
- InterMetro Industries Corporation
- MacArthur Place Hotel, Sonoma
- National Chicken Council
- The Spice House.com
- True FoodService Equipment, Inc.
- The Vollrath Company, LLC
- Vulcan-Hart, a division of the ITW Food Equipment Group LLC
- Wisconsin Milk Marketing Board, Inc.

Contents

Calculating Measurements 49

Chapter 3

Converting Measurements and Scaling Recipes 85

Chapter 4

Contents

Chapter 6 (continued)

Calculating Revenue and Expenses 169

Chapter 7

Contents

Interactive CD-ROM Contents

- *Quick Quizzes®*
- *Illustrated Glossary*
- *Master Math™ Applications*
- *Media Clips*

- *Flash Cards*
- *Forms and Tables*
- *ATPeResources.com*

CD-ROM Features

The Interactive CD-ROM located in the back of this text-workbook is a self-study aid that reinforces the content in Culinary Math Principles and Applications. This Windows® compatible CD-ROM includes the following interactive learning tools:

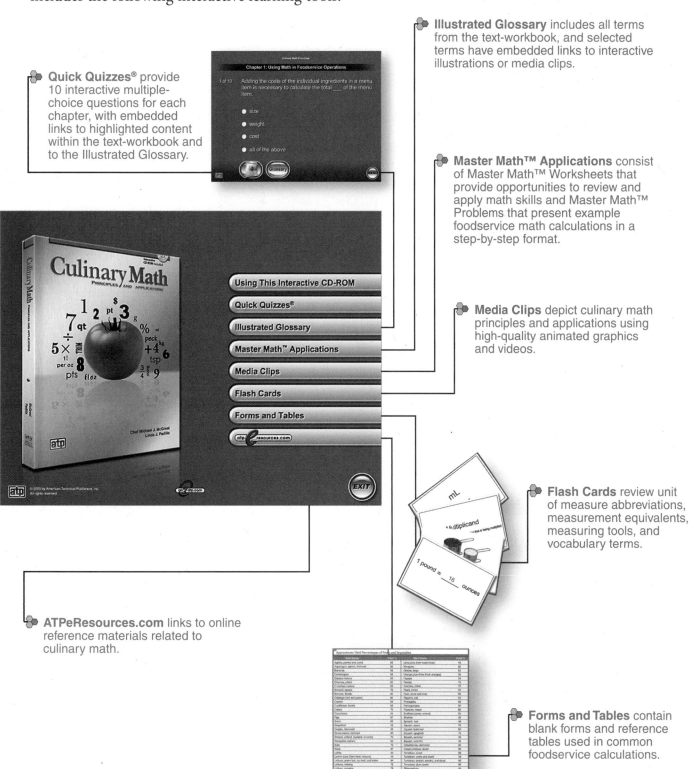

Illustrated Glossary includes all terms from the text-workbook, and selected terms have embedded links to interactive illustrations or media clips.

Quick Quizzes® provide 10 interactive multiple-choice questions for each chapter, with embedded links to highlighted content within the text-workbook and to the Illustrated Glossary.

Master Math™ Applications consist of Master Math™ Worksheets that provide opportunities to review and apply math skills and Master Math™ Problems that present example foodservice math calculations in a step-by-step format.

Media Clips depict culinary math principles and applications using high-quality animated graphics and videos.

Flash Cards review unit of measure abbreviations, measurement equivalents, measuring tools, and vocabulary terms.

ATPeResources.com links to online reference materials related to culinary math.

Forms and Tables contain blank forms and reference tables used in common foodservice calculations.

Introduction

Culinary Math Principles and Applications describes and illustrates how and why foodservice workers use math every day in the professional kitchen. This text-workbook integrates math skills within the culinary arts in an easy-to-follow style that helps learners grasp key principles and applications. Each chapter is divided into "Sections" by subtopic, and each section ends with a "Checkpoint" that is comprised of short-answer review questions requiring the application of key principles.

Checkpoint Answers, a Chapter Summary, and several pages of Math Exercises are provided at the end of each chapter. The Appendix contains a listing of math formulas, reference tables, and blank forms for use when performing foodservice math calculations.

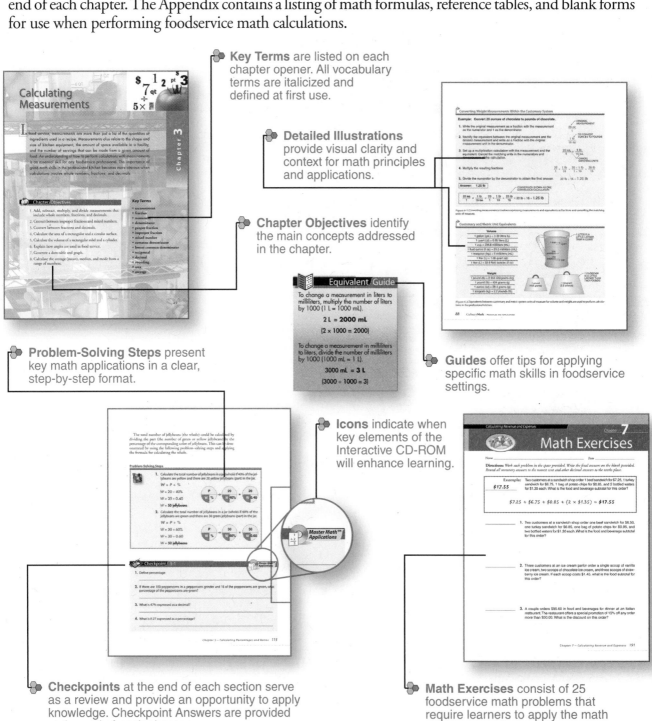

Key Terms are listed on each chapter opener. All vocabulary terms are italicized and defined at first use.

Detailed Illustrations provide visual clarity and context for math principles and applications.

Chapter Objectives identify the main concepts addressed in the chapter.

Guides offer tips for applying specific math skills in foodservice settings.

Problem-Solving Steps present key math applications in a clear, step-by-step format.

Icons indicate when key elements of the Interactive CD-ROM will enhance learning.

Checkpoints at the end of each section serve as a review and provide an opportunity to apply knowledge. Checkpoint Answers are provided at the end of each chapter.

Math Exercises consist of 25 foodservice math problems that require learners to apply the math skills covered in the chapter.

Using Math in Foodservice Operations

Math plays a very important role in the success of every type of food-service operation. Math is used to perform calculations related to the menu such as ordering food and supplies, preparing recipes, servicing customers, and managing general finances. Math skills are required at all levels of employment from servers and cooks to chefs and managers. Therefore, it is important to understand the many ways in which math is used in foodservice operations.

Chapter Objectives

1. Describe how math is used everyday in a foodservice operation both inside and outside of the kitchen.
2. Explain why the success of a foodservice operation depends on the use of math skills.
3. Explain how to add and subtract whole numbers.
4. Explain how to multiply and divide whole numbers.

Key Terms

- **whole number**
- **addition**
- **sum**
- **subtraction**
- **multiplication**
- **product**
- **multiplicand**
- **multiplier**
- **division**
- **divisor**
- **dividend**
- **quotient**

HOW MATH IS USED IN FOOD SERVICE

A career in food service can be exciting and rewarding. However, many people often make the mistake of thinking that working in food service is simply fun and entertaining because of the glamour associated with celebrity chefs on television. While cooking shows are entertaining, they rarely focus on the less glamorous side of the business, especially the physically demanding nature of the work. The truth is that the winning personality of a celebrity chef will only take you so far in the foodservice industry. The path to success requires a strong culinary background and a working knowledge of basic business skills. The most fundamental business skill is a solid understanding of basic math.

A foodservice operation cannot succeed unless employees, managers, and owners have solid math skills. Math is used in virtually all areas of the operation such as the following:

- pricing menu items
- ordering food and supplies
- measuring recipe ingredients
- preparing food
- serving food
- storing food and supplies
- processing money
- scheduling and payroll
- tracking income and expenses

Pricing Menu Items

Foodservice operations, such as restaurants, use menus to show customers the items that are available and the prices of those items. **See Figure 1-1.** Math skills are required when determining the price of the items on a menu. A foodservice operation only earns money if the operation charges more money for the items that it sells than the amount of money it spends to produce those items. Having menu items priced appropriately and selling enough of them are primary factors in determining whether a foodservice operation will succeed.

Menu Prices

LET'S GO BISTRO

STARTERS & SOUPS

Cajun-Style Crab Cakes with Honey-Jalapeño Sauce $6
   Premium lump crabmeat, Cajun trinity, panko breadcrumbs.
New Orleans BBQ Shrimp.. $8
   Spicy sautéed shrimp in Worcestershire-butter sauce served with a cheddar risotto cake.
Caramelized Onion Tartlets $5
   Crispy filo dough cups filled with a sweet onion filling topped with Emmenthaler cheese.
Potato-Leek Soup with Wild Rice and Cheddar Cheese $4
   Our signature soup.
"Ruth's" Italian Style Chicken Noodle Soup $4
   Full flavored chicken broth with a touch of tomato, veggies, and egg noodles. Topped with a sprinkling of Parmigianno-Reggiano cheese.

Figure 1-1. Math is used to determine the price that customers are charged for each of the items on a menu.

The math involved in determining the price of an item on a menu starts with adding the cost of the ingredients in the recipe. For example, calculating the cost to prepare a pizza can be very simple in the case of a cafeteria owner who purchases premade frozen pizzas. The cost is simply the amount that the owner pays the supplier of the frozen pizzas. However, if a restaurant makes a pizza from scratch, the cost of the pizza is calculated by adding the individual costs of all the ingredients used in the recipe. For example, the ingredients in a cheese pizza would include the dough, sauce, and cheese. **See Figure 1-2.**

Another factor to consider when determining the price of a menu item is the labor involved. In the case of the cafeteria serving premade frozen pizza, the labor is limited to having an employee unwrap the pizza and bake it in an oven. With a restaurant pizza made from scratch, employees are needed to mix, knead, and portion the dough; shape the crust; prepare and portion the toppings; assemble the pizza, and bake it. The price of pizza charged by the restaurant will be higher than the price charged by the cafeteria because of the additional labor costs.

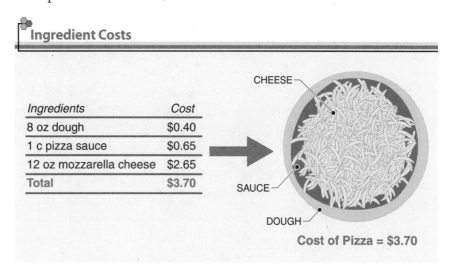

Ingredient Costs

Ingredients	Cost
8 oz dough	$0.40
1 c pizza sauce	$0.65
12 oz mozzarella cheese	$2.65
Total	$3.70

CHEESE
SAUCE
DOUGH

Cost of Pizza = $3.70

Figure 1-2. The individual cost of each ingredient in a recipe must be added to calculate the total cost of a menu item made from scratch.

Ordering Food and Supplies

Before ordering food and supplies, foodservice employees evaluate the prices charged for similar products from different suppliers in order to keep expenses to a minimum. Math is involved in making these evaluations because the process is not always as simple as comparing one supplier's price to another. Often, adjustments must be made to account for the differences in the way products are packaged.

For example, if two suppliers sell identical 50-pound bags of flour, deciding which to buy from is as simple as identifying the lowest price. However, if one supplier sells flour in a 25-pound bag and another supplier sells the same brand of flour in a 50-pound bag, calculations are required to determine which one offers the better price. Math is used to calculate the price of the flour per pound as opposed to the price per bag so that the prices can be accurately compared. **See Figure 1-3.** If a foodservice operation uses thousands of pounds of flour every year, even a small amount of savings on a per pound basis can reduce expenses significantly.

PRICE PER BAG

PRICE PER POUND

Figure 1-3. Math skills are used to determine which packaging option represents the better price.

Measuring Recipe Ingredients

Math is used everyday in the professional kitchen, especially when adjusting the quantities of individual ingredients in recipes. For example, a cook working on a weekday might be responsible for making enough chicken noodle soup to feed 50 people. However, on weekends when the restaurant might be busier than normal, the chef might tell the cook to make enough chicken noodle soup to serve 100 people. The cook will need to know how to adjust the quantities of each ingredient used in the recipe in order to prepare 100 servings instead of 50 servings. **See Figure 1-4.**

When the ingredients in a recipe are increased or decreased, it is often necessary to convert the units used to measure an ingredient. For example, if a typical recipe calls for 3 tablespoons of milk and 10 times the recipe needs to be made, 30 tablespoons of milk (3 tablespoons × 10 = 30 tablespoons) would need to be measured. It would take a long time to measure out 30 tablespoons of milk one tablespoon at a time. However, the measurement could be converted from tablespoons to a larger unit of measure, such as cups, to measure the milk more efficiently. In order to do this, the cook must know how many tablespoons are in a cup and then calculate the number of cups needed for the larger recipe.

Calculating Ingredient Quantities

Chicken Noodle Soup

Ingredients	50 servings	100 servings
Chicken Stock	3 gal.	6 gal.
Carrots, diced	3 lb	6 lb
Celery, diced	2 lb	4 lb
Onions, diced	2 lb	
Egg Noodles, cooked	2 lb	
Salt	2 Tbsp	
Black Pepper	4 tsp	

Method
1. Bring chicken stock to a boil.
2. Add carrots, celery, and onions

Figure 1-4. Math is used to recalculate the amount of ingredients to be used in a recipe when the number of servings is changed.

Preparing Food

Many items used in restaurants, such as chicken, may need to be cut up or broken down in some way before being used in a recipe. The amount of a product that cannot be used (waste) must be taken into account when ordering the product. For example, if a recipe calls for 10 pounds of boneless-skinless diced chicken, more than 10 pounds of whole chickens will be needed to make the recipe. This is because whole chickens contain skin and bones that will not be used in the recipe.

The amount of boneless-skinless, diced chicken produced from cutting up an average-sized chicken must be determined before calculating the amount of whole chickens that will need to be ordered. **See Figure 1-5.** In addition, chicken may be used in several other recipes prepared by the foodservice operation. Therefore, the quantity of chicken required for all recipes being prepared must be added together to determine the total amount of chicken that should be ordered at a given time.

Media Clips — Accounting for Food Waste

Accounting for Food Waste

Whole Chicken   Diced Boneless-Skinless Meat   Skin and Bones (Waste)

Figure 1-5. The amount of waste generated when cutting up or breaking down a food item must be taken into account when ordering food.

Serving Food

One of the most common uses of math in a foodservice operation is to determine how much food should be prepared to serve a group of people at an event, such as a wedding. If 120 people attend an event and the kitchen runs out of food after serving 100 people, 20 people will not be served and the event will not be successful. The opposite extreme is also possible. If, after serving 120 people, there is enough food left over to feed an additional 20 people, money has been wasted. Using math skills to calculate the amount of food to prepare is essential. Guessing will almost certainly result in costly errors.

Storing Food and Supplies

Foodservice operations store enough food and supplies to make sure that the operation does not run out of the things needed to serve customers. Before orders for additional food and supplies are placed, the amount of items already in storage should be counted. Many foodservice operations will have an inventory checklist that lists the amount of each item that should be in storage. After adding how much of each particular item is already in storage, an employee can then compare those amounts to the amounts provided on the checklist and decide which items need to be ordered. **See Figure 1-6.** This process provides a reliable inventory of food and supplies to be on hand for daily use.

Inventory Checklist			
Item	Desired Amount in Storage	Amount Stored	Amount to Order
Spaghetti Noodles	5 Cases	2	3
Macaroni Noodles	4 Cases	3	1
Tomato Sauce	3 Cases	1	2
Tomato Paste	2 Cases	0	2
Bread Crumbs	5 Bags	4	1
Sea Salt, Fine	8 Boxes	5	3
Olive Oil, Extra Virgin	12 Bottles	12	0

Figure 1-6. The amount of food and supplies in storage should be counted before placing additional orders.

Processing Money

Math skills are used to process money at all levels of a foodservice operation. Money constantly comes into a foodservice operation when customers pay their bills. It constantly goes out of a foodservice operation to pay for expenses such as food, payroll, and taxes. Managers and owners closely monitor the flow of money into and out of the operation. For a foodservice operation to be successful, the amount of money coming into the operation must exceed the amount of money going out of the operation.

Employees, such as servers and cashiers, must provide customers with accurate checks for the menu items ordered. They must also provide the right amount of change after payment has been received. **See Figure 1-7.** Managers are usually responsible for adding all of the sales in a foodservice operation and making sure that the total amount of money received matches the total amount of sales recorded.

Managers must also keep track of and pay bills on time. Late payments can result in late fees and bad credit ratings. Poor skills in money management can lead to some providers refusing to continue to do business with a foodservice operation.

Figure 1-7. Servers use math to ensure customers receive accurate checks for the menu items ordered and the proper amount of change after payment is received.

Scheduling and Payroll

Managers spend a great deal of effort determining how many employees are needed during a given period of time. Calculating the appropriate number of employees needed for each shift or day of operation is based on a number of factors. These factors include, but are not limited to, the time of year, predicted sales volume, number of reservations, and employees scheduled to take vacation.

Managers are also responsible for processing payroll to ensure employees receive accurate paychecks in a timely manner. This involves adding the hours worked by each employee and multiplying the hours worked by the appropriate hourly wage. If an employee works overtime, the manager must make adjustments to ensure that the employee is paid correctly.

Tracking Income and Expenses

Most foodservice operations use computerized systems to track income and expenses on a continuous basis. These computerized systems can generate reports that show how a foodservice operation is doing from a financial standpoint at a given point in time. Math skills are necessary for understanding how to categorize different transactions and to make sure they are entered into the system properly. Math skills are also necessary for spotting errors and determining if any action needs to be taken. If financial information is not monitored routinely, a foodservice operation can fail quickly.

Checkpoint 1-1

1. Why is it important to know the cost of each ingredient in a recipe before deciding how much a customer will be charged for that menu item?

_____

_____

2. Why do foodservice operations evaluate how much different suppliers charge for similar products?

_____

_____

3. If a recipe calls for 50 pounds of peeled and diced potatoes, why is more than one 50-pound case of whole, fresh potatoes needed to prepare the recipe?

_____

_____

4. Why is adding the total amount of money that comes into a foodservice operation and the total amount of money that goes out of a foodservice operation important?

_____

_____

PERFORMING BASIC MATH CALCULATIONS

The math used in a foodservice operation involves adding, subtracting, multiplying, and dividing whole numbers. A *whole number* is a number that is used for counting, such as 0, 1, 20, or 100. For example, the number 3 is a whole number, while 3½ and 3.5 are not.

Whole numbers are often used to represent the quantity of something, such as the number of customers on a reservation, the number of cans of ketchup in the supply room, or the number of potatoes that need to be peeled. Whole numbers with several digits are made easier to read by separating the number with commas into sets of three digits each. Each digit in a set occupies a different place value. **See Figure 1-8.**

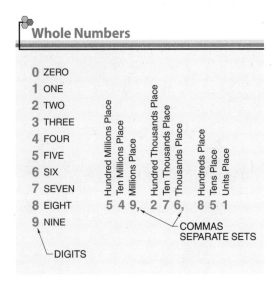

Whole Numbers

0	ZERO
1	ONE
2	TWO
3	THREE
4	FOUR
5	FIVE
6	SIX
7	SEVEN
8	EIGHT
9	NINE

Hundred Millions Place
Ten Millions Place
Millions Place
Hundred Thousands Place
Ten Thousands Place
Thousands Place
Hundreds Place
Tens Place
Units Place

5 4 9, 2 7 6, 8 5 1

COMMAS SEPARATE SETS

DIGITS

Figure 1-8. Whole numbers are separated into sets and places.

Adding Whole Numbers

Adding whole numbers requires an understanding of addition. Addition is used when a number, count, or total is increased. *Addition* is the process of combining two or more numbers into a single number to find the sum. A *sum* is the number that is produced as the result of addition. For example, a server in a banquet hall may be responsible for polishing one glass for every person expected at a party. If there are three groups coming to the party and the first group has 138 people, the second group has 56 people, and the third group has 34 people, the server will need to add up the total number of people to figure out how many glasses to polish. The total number of people coming to the party is 228 (138 people + 56 people + 34 people = 228 people). Since each person needs one glass, the server will need to polish 228 glasses. When adding whole numbers with more than one digit, numbers in the units column are added first, then the tens column, then the hundreds column, and so on. **See Figure 1-9.**

Subtracting Whole Numbers

Subtracting whole numbers requires an understanding of subtraction. Subtraction is used whenever a number, count, or total is decreased. *Subtraction* is the process of taking one number from another number to find the difference.

For example, what if a customer that booked a banquet for 155 people notifies the banquet manager that 47 of the people invited will not be coming to the banquet? By subtracting the number of people who will not be coming from the original number expected, the banquet manager can calculate the total number of people that will attend the banquet. The total number of people attending is 108 (155 people − 47 people = 108 people). **See Figure 1-10.**

Charlie Trotter's

Adding Whole Numbers

Example: A party will be attended by 3 groups. Group 1 has 138 people, Group 2 has 56 people, and Group 3 has 34 people. What is the total number of people attending the party?

1. Arrange numbers vertically in columns.

2. Add the digits in the units column (8 + 6 + 4 = 18). Record the 8 and carry the 1 to the tens column.

3. Add the digits in the tens column (1 + 3 + 5 + 3 = 12). Record the 2 and carry the 1 to the hundreds column.

4. Add the digits in the hundreds column (1 + 1 = 2).

Answer: 228 people

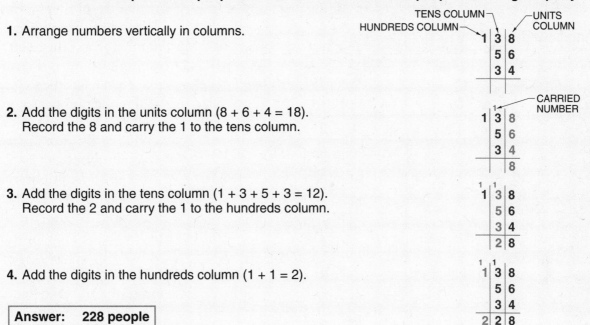

Figure 1-9. Addition can be used in foodservice operations to determine the number of people attending a party.

Subtracting Whole Numbers

Example: A banquet originally planned for 155 guests has received cancellations from 47 of the expected guests. How many guests are still expected for the banquet?

1. Arrange numbers vertically in columns.

2. Borrow "10" from the tens column and add that 10 to the units column. Subtract the digits in the units column (15 – 7 = 8). Record the 8.

3. Subtract the digits in the tens column (4 – 4 = 0). Record the 0.

4. Subtract the digits in the hundreds column (1 – 0 = 1).

Answer: 108 guests

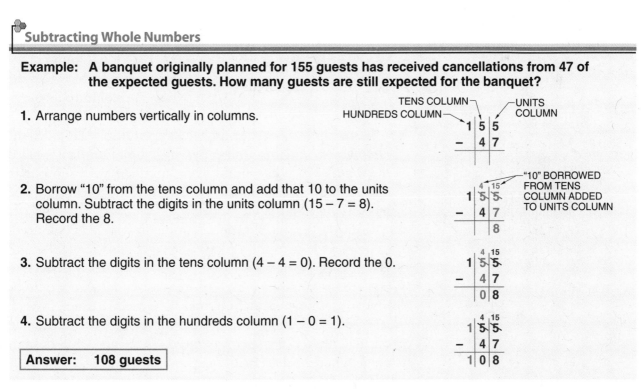

Figure 1-10. Subtraction can be used in foodservice operations to calculate the number of guests still attending a banquet when cancellations are received.

Multiplying Whole Numbers

Multiplying whole numbers requires an understanding of multiplication. Multiplication is often used as a shortcut for addition. *Multiplication* is the process of adding one number to itself any number of times to find the product. A *product* is the number that is the result of multiplication. For example, $2 \times 4 = 8$ is the same as adding the number four to itself two times ($4 + 4 = 8$) or adding the number two to itself four times ($2 + 2 + 2 + 2 = 8$). Multiplication tables can be used to help learn and memorize the products of small whole numbers. For example, to find the product of 4×6, locate the number four in the row (or column) at the top of the multiplication table and the number six in the column (or row) on the left side of the multiplication table. The product (24) of 4×6 is found in the boxes on the table where the rows and columns intersect. **See Figure 1-11.**

Multiplication Table

1	2	3	4	5	6	7	8	9	10	11	12
2	4	6	8	10	12	14	16	18	20	22	24
3	6	9	12	15	18	21	24	27	30	33	36
4	8	12	16	20	24	28	32	36	40	44	48
5	10	15	20	25	30	35	40	45	50	55	60
6	12	18	24	30	36	42	48	54	60	66	72
7	14	21	28	35	42	49	56	63	70	77	84
8	16	24	32	40	48	56	64	72	80	88	96
9	18	27	36	45	54	63	72	81	90	99	108
10	20	30	40	50	60	70	80	90	100	110	120
11	22	33	44	55	66	77	88	99	110	121	132
12	24	36	48	60	72	84	96	108	120	132	144

THE PRODUCT OF BOTH (4×6) AND (6×4) IS 24

Figure 1-11. Multiplication tables can be used to help learn and memorize products of small, whole numbers.

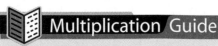

Multiplication Guide

Step 1: Multiply the units digit of the multiplicand by the multiplier ($6 \times 8 = 48$). Record the 8 and carry the 4.

$$\begin{array}{r} {}^{4}16 \\ \times\ 8 \\ \hline 8 \end{array}$$

Step 2: Multiply the tens digit of the multiplicand by the multiplier ($1 \times 8 = 8$) and add the carried number (4). Next, record the result ($8 + 4 = 12$).

$$\begin{array}{r} {}^{4}16 \\ \times\ 8 \\ \hline 128 \end{array}$$

In multiplication, the number being multiplied is called the *multiplicand* and the number by which it is multiplied is called the *multiplier*. When multiplying numbers with more than one digit, every digit in the multiplicand must be multiplied by every digit in the multiplier.

Consider a multiplication example in which a cook needs to prepare desserts for a party where people sit at tables of 8 people each. If the cook has to calculate how many desserts to prepare for 16 tables with 8 people at each table, it would be time consuming to add the number 8 to itself 16 times. Instead, using multiplication as a shortcut for addition, the cook can calculate that 128 desserts are needed (16 tables × 8 desserts per table = 128 desserts). If each table were to seat 12 people instead of 8, the server must prepare 192 desserts (16 tables × 12 desserts per table = 192 desserts). **See Figure 1-12.**

Example: A cook must prepare 12 desserts for each table at a party. How many desserts must be prepared for a party with 16 tables?

1. Arrange numbers vertically in columns as a multiplicand and a multiplier.

2. Multiply the multiplicand by the digit in the units column of the multiplier (16 × 2 = 32). Record the product aligned with the units column.

3. Multiply the multiplicand by the digit in the tens column of the multiplier (1 × 16 = 16). Record the product on the next line aligned with the tens column.

4. Add the first product from Step 2 to the second product from Step 3 to calculate the final answer.

Answer: 192 desserts

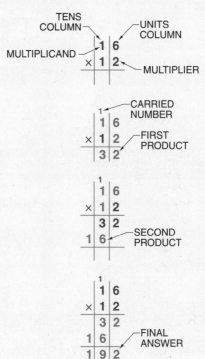

Figure 1-12. Multiplication can be used in foodservice operations when calculating how many desserts need to be prepared for a party.

Dividing Whole Numbers

Dividing whole numbers requires an understanding of division. Just as multiplication is a shortcut for addition, division is a shortcut for subtraction. *Division* is the process of counting how many times one number can go into another number. For example, dividing 8 by 2, which can be written 8 ÷ 2, is a shortcut for calculating how many times 2 needs to be subtracted from 8 to arrive at 0. The answer is 4 times (8 − 2 = 6 and 6 − 2 = 4 and 4 − 2 = 2 and 2 − 2 = 0).

Division can also be thought of as the opposite of multiplication. For example, the multiplication table shows that 7 × 8 = 56. It is also true that 56 ÷ 8 = 7 and that 56 ÷ 7 = 8. In division, the *dividend* is the number being divided, and the *divisor* is the number the dividend is "divided by." In the equation 56 ÷ 7 = 8, 56 is the dividend, 7 is the divisor, and 8 is the quotient. A *quotient* is the number that is the result of division. Any of four different symbols may be used to indicate division (÷, ⁀‾, /, and —).

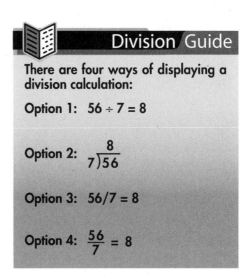

Division Guide

There are four ways of displaying a division calculation:

Option 1: 56 ÷ 7 = 8

Option 2: $7\overline{)56}$ with quotient 8

Option 3: 56/7 = 8

Option 4: $\frac{56}{7} = 8$

Division is used everyday in a foodservice operation. For example, if a customer books a party for 96 people and the banquet hall uses tables that seat 4 people each, the banquet manager must determine how many tables will be needed for the party. The banquet manager could start with 96 and keep subtracting 4 until getting down to zero and then count up the number of subtractions, but that would take a long time. Instead, as a shortcut for subtraction, the banquet manager can use division to determine that 24 tables are needed (96 people ÷ 4 people per table = 24 tables). **See Figure 1-13.**

Dividing Whole Numbers

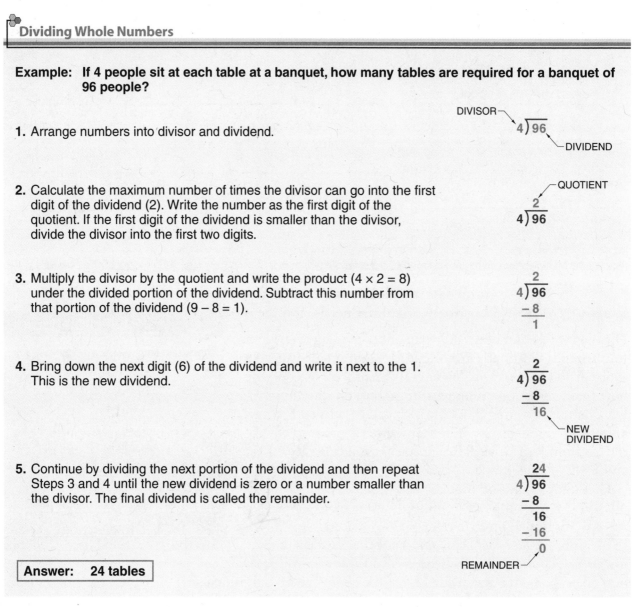

Example: If 4 people sit at each table at a banquet, how many tables are required for a banquet of 96 people?

1. Arrange numbers into divisor and dividend.

2. Calculate the maximum number of times the divisor can go into the first digit of the dividend (2). Write the number as the first digit of the quotient. If the first digit of the dividend is smaller than the divisor, divide the divisor into the first two digits.

3. Multiply the divisor by the quotient and write the product (4 × 2 = 8) under the divided portion of the dividend. Subtract this number from that portion of the dividend (9 − 8 = 1).

4. Bring down the next digit (6) of the dividend and write it next to the 1. This is the new dividend.

5. Continue by dividing the next portion of the dividend and then repeat Steps 3 and 4 until the new dividend is zero or a number smaller than the divisor. The final dividend is called the remainder.

Answer: 24 tables

Figure 1-13. Division can be used to calculate how many tables are needed to seat a given number of people.

1. Define whole number.

_____

2. If a banquet hall has one party with 50 people, a second party with 75 people, and a third party with 115 people, how many total people are there?

_____

3. If 50 chicken breasts need to be grilled for a party and only 12 have been grilled so far, how many more need to be grilled?

_____

4. How many total pieces of silverware are needed at a table with 8 seats if each seat requires 2 forks, 1 knife, and 1 spoon?

_____

_____

5. If the bill for a party is $500 and it is to be split evenly by 4 customers, how much must each customer pay?

_____

Chapter 1 Summary

Quick Quiz® Chapter 1

Flash Cards

Math is used routinely in calculations involving food preparation, planning for customer needs, and the processing of money. Math is essential for ensuring the overall success of a foodservice operation. The majority of the math used in food service involves addition, subtraction, multiplication, and division. Multiplication is a shortcut for addition and division is a shortcut for subtraction. A foodservice operation cannot succeed if the management and staff do not have solid math skills.

Checkpoint Answers

Checkpoint 1-1

1. A foodservice operation must charge more for the menu item than it costs to prepare it in order to earn money.

2. By determining which supplier offers the best price, the foodservice operation can keep expenses to a minimum.

3. Waste will be generated when the fresh potatoes are peeled.

4. The amount of money coming into a foodservice operation must be more than the amount of money going out in order for the operation to be successful.

Checkpoint 1-2

1. A whole number is a number that is used for counting, such as 0, 1, 20, or 100.

2. 240 people   (50 people + 75 people + 115 people = 240 people)

3. 38 chicken breasts (50 chicken breasts – 12 chicken breasts = 38 chicken breasts)

4. 32 pieces (2 forks + 1 knife + 1 spoon = 4 pieces; and 4 pieces per seat × 8 seats = 32 pieces)

5. $125 per customer   ($500 ÷ 4 customers = $125 per customer)

Measuring in the Professional Kitchen

Foodservice operations must be able to produce food that is consistent day after day in order to be successful. The accurate measurement of the ingredients used in recipes is key to maintaining that consistency. Every employee must understand the standards used for measurement and also be able to use the wide variety of measurement tools found in the professional kitchen. Math skills are required to perform these important measurements.

Chapter Objectives

1. Explain the role of standardized measures in the success of foodservice operations.

2. Identify the two systems of measurement used by foodservice operations.

3. List the common units of measure for volume and for weight that are used in the professional kitchen.

4. Define measurement equivalents and explain how they are used in a professional kitchen.

5. Change a measurement given in one unit of measure to an equivalent measurement in another unit of measure.

6. Explain how to use professional kitchen tools to measure volume and weight.

7. Explain how time, temperature, and distance are measured in a foodservice operation.

Key Terms

- **standardized recipe**
- **unit of measure**
- **weight**
- **volume**
- **density**
- **measurement equivalent**
- **measuring spoon**
- **dry measuring cup**
- **liquid measuring cup**
- **ladle**
- **portion-controlled scoop**
- **mechanical scale**
- **digital scale**
- **balance scale**

USING STANDARDIZED MEASURES

One of the most significant challenges for any foodservice operation is to produce food that is consistent. Consistency is important because customers expect food from a particular foodservice operation to look and taste the same every time it is ordered. Since the same cook does not always prepare the same items each day, foodservice operations use standardized recipes. A *standardized recipe* is a list of ingredients, ingredient amounts, and procedural steps for preparing a specific quantity of a food item. **See Figure 2-1.**

When used properly, standardized recipes help to ensure consistent results regardless of the individual preparing the food. Consider a restaurant known for its signature barbecue sauce. What if the chef is the only person who knows how to prepare the signature sauce and is on vacation when another batch of the sauce needs to be made? The restaurant would not be able to offer its signature dishes or another cook would need to guess at the ingredients and method for preparing the sauce.

By guessing at the ingredients, the cook is likely to produce a sauce that does not taste the same. This situation would most likely result in dissatisfied customers and a decline in business. Having a standardized recipe for the signature barbecue sauce would avoid this problem.

A standardized recipe includes a list of ingredients and uses standard units of measure to represent a specific amount of each ingredient. A *unit of measure* is a fixed quantity that is widely accepted as a standard of measurement. *Volume* is a measurement of the physical space a substance occupies. *Weight* is a measurement of the heaviness of a substance. Units of measure are also used in standardized recipes to indicate time, temperature, and distance (size).

Foodservice operations commonly use two different measurement systems to measure food and beverage products. The customary system uses the ounce (oz) and the pound (lb) to measure weight, and the teaspoon (tsp), tablespoon (tbsp), fluid ounce (fl oz), cup (c), pint (pt), quart (qt), and gallon (gal.) to measure volume. The inch (in.) and foot (ft) measure distance. The metric system uses the gram (g) to

*The Vollrath Co., LLC

Standardized Recipe

Shrimp Creole		
Yield: 25 Servings	**Cooking Temperature:** 175°F	
Portion Size: 9 oz	**Cooking Time:** 20 min	

Amount	Ingredients	Procedure
8 lb	shrimp	1. Cook shrimp by steaming.
3 qt	prepared creole sauce	2. Bring creole sauce to a simmer in saucepot.
to taste	salt and pepper	3. Add cooked shrimp. Blend gently into sauce and season to taste.
3 lb	prepared white rice	4. Place 2 oz preprared rice on each serving plate.
		5. Dish shrimp into casserole with 6 oz ladle.

Figure 2-1. A standardized recipe is a list of ingredients, ingredient amounts, and procedural steps for preparing a specific quantity of a food item.

measure weight, the liter (L) to measure volume, and the meter (m) to measure distance. To measure temperature, the customary system uses degrees Fahrenheit (°F) and the metric system uses degrees Celsius (°C). **See Figure 2-2.**

Common Food Service Units of Measure

Volume Units		Temperature Units		Weight Units		Distance Units	
Customary System		**Customary System**		**Customary System**		**Customary System**	
Unit	*Abbreviation*	*Unit*	*Abbreviation*	*Unit*	*Abbreviation*	*Unit*	*Abbreviation*
teaspoon	tsp	degrees Fahrenheit	°F	ounce	oz	inch	in.
tablespoon	tbsp			pound	lb or #	foot	ft
fluid ounce	fl oz						
cup	c	**Metric System**		**Metric System**		**Metric System**	
pint	pt	*Unit*	*Abbreviation*	*Unit*	*Abbreviation*	*Unit*	*Abbreviation*
quart	qt	degrees Celsius	°C	milligram	mg	millimeter	mm
gallon	gal.			gram	g	centimeter	cm
				kilogram	kg	meter	m
Metric System							
Unit	*Abbreviation*						
liter	L						
milliliter	mL						

Figure 2-2. Professional kitchens use measurement units of both the customary system and the metric system.

Checkpoint 2-1

1. Explain why consistency is important in the foodservice industry.

_____

_____

2. Define standardized recipe.

_____

_____

3. Define unit of measure.

_____

_____

4. Explain the difference between volume and weight.

_____

_____

MEASURING VOLUME

In the professional kitchen, the most common volume measurement units and their abbreviations are the milliliter (mL), teaspoon (tsp), tablespoon (tbsp), fluid ounce (fl oz), cup (c), pint (pt), quart (qt), liter (L), and gallon (gal.). The tools used in the professional kitchen to measure volume are measuring spoons, dry measuring cups, liquid measuring cups, ladles, and portion-controlled scoops.

Ounces Versus Fluid Ounces

Before learning to measure ingredients, it is important to understand the difference between an ounce (oz) and a fluid ounce (fl oz). An ounce is a measurement of weight and a fluid ounce is a measurement of volume. Failure to understand this distinction often leads to costly mistakes. Some ingredients, when measured by volume in fluid ounces, will also weigh the same amount in ounces. For example, 8 fluid ounces of water weighs almost exactly 8 ounces. However, 8 fluid ounces of all-purpose flour only weighs about 4½ ounces.

Carlisle FoodService Products

The reason that 8 fluid ounces of flour weighs less than 8 ounces is that flour has a lower density than water. *Density* is the measure of how much a given volume of a substance weighs. It is important to remember the few ingredients that can be measured in either fluid ounces or ounces without affecting a recipe. These ingredients include water and substances with densities very close to water such as alcohol, juices, vinegar, oil, milk, butter, eggs, and granulated sugar.

Other food ingredients have a higher density than water and will, therefore, measure more by weight in ounces than by volume in fluid ounces. For example, 8 fluid ounces of honey measured by volume will actually weigh about 12 ounces. Some ingredients frequently thought to be similar in density to water, but are not, include honey, molasses, and various types of syrup. **See Figure 2-3.**

If all standardized recipes were carefully written to distinguish between ounces and fluid ounces this issue would not be as important. Unfortunately, many recipes are not written clearly and require the cook to pay close attention and to exercise proper judgment. When in doubt about recipe amounts, it is best to ask for clarification before measuring.

Volume Units and Equivalents

The largest volume unit of measure used in the professional kitchen is the gallon and the smallest used is the milliliter. In addition to learning which units of measure are smaller or larger than others, foodservice employees must also understand the concept of measurement equivalents.

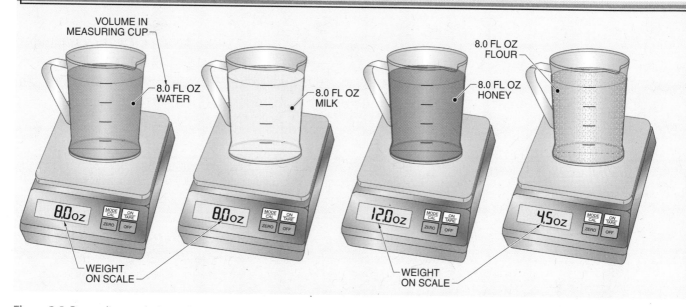

Figure 2-3. Depending on the ingredient being measured, a measurement in fluid ounces will not always be the same as a measurement in ounces.

A *measurement equivalent* is the amount of one unit of measure that is equal to another unit of measure. For example, 1 gallon is the equivalent of 4 quarts, just as 1 dollar is the equivalent of 4 quarters. Other ways of stating this equivalency are to say that "there are 4 quarts in a gallon," "there are 4 quarts per gallon," or "4 quarts make up 1 gallon."

Foodservice employees should memorize the basic measurement equivalents for volume and, when necessary, be able to calculate the equivalents between any two volume units of measure. **See Figure 2-4.** For example, a chef would expect a cook to know that 1 quart is equal to 2 pints and that there are 8 fluid ounces in 1 cup. However, even an experienced chef may need to pull out a pencil and paper or a calculator to figure out how many teaspoons are in a gallon. The following basic volume measurement equivalents should be memorized:

Media Clips — Ounces vs. Fluid Ounces

- 4 qt = 1 gal.
- 2 pt = 1 qt
- 2 c = 1 pt
- 8 fl oz = 1 c
- 2 tbsp = 1 fl oz
- 3 tsp = 1 tbsp

Media Clips — Volume Equivalents

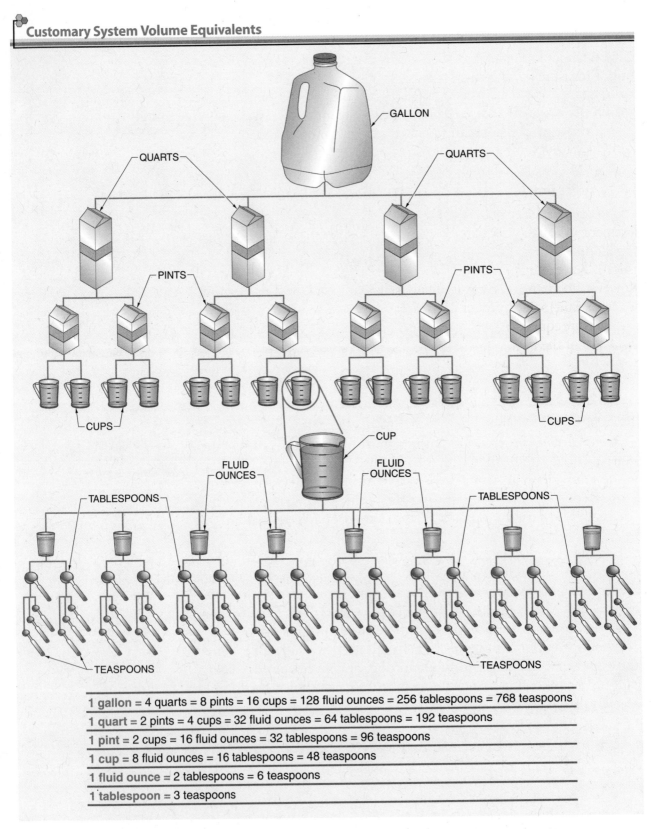

1 gallon = 4 quarts = 8 pints = 16 cups = 128 fluid ounces = 256 tablespoons = 768 teaspoons

1 quart = 2 pints = 4 cups = 32 fluid ounces = 64 tablespoons = 192 teaspoons

1 pint = 2 cups = 16 fluid ounces = 32 tablespoons = 96 teaspoons

1 cup = 8 fluid ounces = 16 tablespoons = 48 teaspoons

1 fluid ounce = 2 tablespoons = 6 teaspoons

1 tablespoon = 3 teaspoons

Figure 2-4. Foodservice employees must understand how different volume units of measure relate to one another.

Calculating Volume Equivalents

Using the basic volume equivalents, it is easy to change a measurement with one unit of measure to an equivalent measurement with a different unit of measure by following these two rules:

- To change from a larger to smaller unit of measure, multiply the number in the original measurement by the number of smaller units that make up the larger unit.
- To change from a smaller to a larger unit, divide the number in the original measurement by the number of smaller units that make up the larger unit.

Consider the basic equivalent 4 quarts = 1 gallon. To change a measurement in gallons to quarts (larger unit to smaller unit), the number of gallons is multiplied by 4 (the number of quarts that make up one gallon). For example, 2 gallons are equivalent to 8 quarts ($2 \times 4 = 8$). However, to change a measurement in quarts to gallons (smaller unit to larger unit), the number of quarts is divided by 4. For example, 12 quarts are equivalent to 3 gallons ($12 \div 4 = 3$). The same process works using any of the basic volume equivalents. **See Figure 2-5.**

Calculating Volume Equivalents

Basic Volume Equivalents	To Change:	Example Calculations	
4 qt = 1 gal.	gallons → quarts (multiply by 4)	2 gal. = 8 qt ($2 \times 4 = 8$)	CHANGING FROM LARGER UNIT TO SMALLER UNIT USES MULTIPLICATION
	quarts → gallons (divide by 4)	12 qt = 3 gal. ($12 \div 4 = 3$)	
2 pt = 1 qt	quarts → pints (multiply by 2)	2 qt = 4 pt ($2 \times 2 = 4$)	
	pints → quarts (divide by 2)	6 pt = 3 qt ($6 \div 2 = 3$)	
2 c = 1 pt	pints → cups (multiply by 2)	8 pt = 16 c ($8 \times 2 = 16$)	
	cups → pints (divide by 2)	4 c = 2 pt ($4 \div 2 = 2$)	CHANGING FROM SMALLER UNIT TO LARGER UNIT USES DIVISION
8 fl oz = 1 c	cups → fluid ounces (multiply by 8)	2 c = 16 fl oz ($2 \times 8 = 16$)	
	fluid ounces → cups (divide by 8)	24 fl oz = 3 c ($24 \div 8 = 3$)	
2 tbsp = 1 fl oz	fluid ounces → tablespoons (multiply by 2)	5 fl oz = 10 tbsp ($5 \times 2 = 10$)	
	tablespoons → fluid ounces (divide by 2)	12 tbsp = 6 fl oz ($12 \div 2 = 6$)	
3 tsp = 1 tbsp	tablespoons → teaspoons (multiply by 3)	4 tbsp = 12 tsp ($4 \times 3 = 12$)	
	teaspoons → tablespoons (divide by 3)	6 tsp = 2 tbsp ($6 \div 3 = 2$)	

Figure 2-5. Basic volume measurement equivalents can be used to calculate equivalent volume measurements with different units of measure.

Other equivalent measurements may need to be calculated in more than one step. For example, if a standardized recipe calls for 8 tablespoons of sugar and the chef instructs the cook to make 4 times the recipe, the cook would need to measure 32 tablespoons (8 tbsp × 4 = 32 tbsp). Since it would take too long to measure 32 tablespoons of sugar individually, the tablespoons are converted to a larger unit of measure such as a cup. Converting 32 tablespoons to 2 cups is done in two steps. Since this conversion is from a smaller unit of measure (tablespoons) to a larger unit of measure (cups), division is used.

First, since 2 tablespoons make up 1 fluid ounce, the measurement in tablespoons is changed to fluid ounces by dividing the number of tablespoons by 2.

32 tbsp = **16 fl oz** (32 ÷ 2 = 16)

Then, since 8 fluid ounces make up 1 cup, the measurement in fluid ounces is changed to cups by dividing the number of fluid ounces by 8.

16 fl oz = **2 c** (16 ÷ 8 = 2)

Volume equivalents within the metric system are easy to remember once the concept of metric prefixes is understood. The major unit of volume in the metric system is the liter. All other metric units for volume contain the word "liter" but also contain a prefix based on multiplying or dividing a liter by factors of 10.

For example, a deciliter (deci-liter) is the equivalent of one-tenth of a liter or, stated another way, there are 10 deciliters in 1 liter. A hectoliter (hecto-liter) is the equivalent of 100 times the amount of a liter (there are 100 liters in 1 hectoliter). **See Figure 2-6.** In the professional kitchen the most common metric volume measurement used besides the liter is the milliliter (1000 milliliters equal 1 liter). Occasionally, it may be necessary to perform calculations based on the equivalents between customary units and metric units, such as calculating how many cups are in a liter. These calculations are discussed in Chapter 4: Converting Measurements and Scaling Recipes.

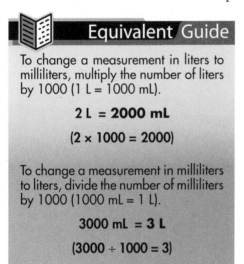

Equivalent Guide

To change a measurement in liters to milliliters, multiply the number of liters by 1000 (1 L = 1000 mL).

2 L = 2000 mL

(2 × 1000 = 2000)

To change a measurement in milliliters to liters, divide the number of milliliters by 1000 (1000 mL = 1 L).

3000 mL = 3 L

(3000 ÷ 1000 = 3)

Metric System Equivalents

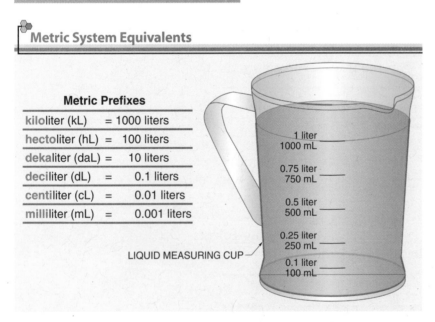

Metric Prefixes

kiloliter (kL)	=	1000 liters
hectoliter (hL)	=	100 liters
dekaliter (daL)	=	10 liters
deciliter (dL)	=	0.1 liters
centiliter (cL)	=	0.01 liters
milliliter (mL)	=	0.001 liters

1 liter
1000 mL

0.75 liter
750 mL

0.5 liter
500 mL

0.25 liter
250 mL

0.1 liter
100 mL

LIQUID MEASURING CUP

Figure 2-6. The metric system uses prefixes to indicate how many times the main unit is increased or decreased by factors of 10.

Tools for Measuring Volume

The most common tools for measuring volume include measuring spoons, dry measuring cups, liquid measuring cups, ladles, and portion-controlled scoops. **See Figure 2-7.** Both dry and liquid ingredients can be measured using volume measurement tools. Regardless of the tool used, ingredients should always be measured as stated in the recipe.

Proper techniques must be followed when using these tools in order to make accurate measurements. If even a slight distinction in recipe wording is overlooked, it can have a major impact on the quality of the final product being prepared. For example, if a recipe calls for "1 cup flour, sifted," measure 1 cup of flour and then sift the flour. However, if a recipe calls for "1 cup sifted flour," sift the flour first and then measure 1 cup of the flour.

Measuring Spoons. A *measuring spoon* is a small volume measurement tool, shaped like a spoon, that is used to measure liquid or dry ingredients. There are usually five spoons, of different sizes, in a set and they are often connected to each other by a ring. The ring prevents one of the spoons from getting lost and ensures that the person using them will always have the five spoons available as needed.

The spoons are designed to measure one-eighth of a teaspoon (⅛ tsp), one-quarter of a teaspoon (¼ tsp), one-half of a teaspoon (½ tsp), one teaspoon (1 tsp), and one tablespoon (1 tbsp). Some recipes may call for a "pinch" or a "dash" of an ingredient. These are not standard units of measure that mean the same to every cook. However, one-eighth of a teaspoon is the most appropriate measure to use when these terms are encountered.

When measuring spoons are used to measure liquid ingredients, the spoons should be filled to the top. When measuring dry ingredients, the spoons should be intentionally overfilled and the ingredient leveled to remove any excess. An ingredient can be leveled by skimming a spatula or the back of a knife across the top of the spoon. Care should be taken not to compress the dry ingredient into the measuring spoon unless directed to do so by the recipe. For example, a recipe may call for 2 tablespoons of brown sugar, firmly packed.

Volume Measurement Tools

Measuring Spoons

Dry Measuring Cups

Liquid Measuring Cups

Ladles

Portion-Controlled Scoops

Carlisle FoodService Products

Figure 2-7. A variety of tools for measuring volume are used in the professional kitchen.

Dry Measuring Cups. A *dry measuring cup* is a volume measurement tool, shaped like a cup, with a short handle used to measure a specific volume of dry ingredients. A dry measuring cup must be filled to the top in order to provide an accurate measurement. Since it is difficult to move a completely filled cup of liquid without spilling some of the contents, liquids should not be measured in dry measuring cups. Dry measuring cups are usually kept in a set consisting of one-quarter of a cup (¼ c), one-third of a cup (⅓ c), one-half of a cup (½ c), and one cup (1 c). Some sets also include two-thirds of a cup (⅔ c) and three-fourths of a cup (¾ c).

Unless a recipe states otherwise, the proper procedure for measuring dry ingredients in a dry measuring cup is to spoon the ingredient into the cup, initially overfilling the cup, and then leveling to remove the excess. Unfortunately, many cooks develop the bad habit of dipping the dry measuring cup directly into the ingredient being measured, such as a bin of powdered sugar, and pressing the cup against the edge of the container to level it. This compresses the ingredient into the cup, resulting in more product being measured than when the proper technique is followed. This inconsistency can adversely affect the final product. Therefore, dry ingredients are often weighed on a scale to obtain more consistent measurements.

Liquid Measuring Cups. A *liquid measuring cup* is a volume measurement tool, shaped like a cup or pitcher, with graduated markings on the side that indicate the volume of a liquid. For example, a one-cup liquid measuring cup may also be marked so that the user can measure one-quarter, one-third, one-half, or two-thirds of a cup. Some liquid measuring cups are also marked to measure teaspoons, tablespoons, fluid ounces, quarts, or gallons.

Liquid measuring cups are the best volume measurement tools for measuring liquid ingredients. The main advantage over other tools is that liquid measuring cups are designed to hold more than the maximum amount to be measured. Therefore, they are never filled to the very top. Because of this extra room, a person is less likely to spill the measured ingredients.

Liquid measuring cups used in the professional kitchen are made of aluminum, stainless steel, or plastic. When measuring, they should always be placed on a level surface and viewed at eye level so that the ingredient is measured accurately. Ingredients should always be poured right up to the marking or line on the measuring cup that indicates the desired amount. **See Figure 2-8.**

Using a Liquid Measuring Cup

**Level Cup
(Accurate Reading)**

**Unlevel Cup
(Inaccurate Reading)**

Figure 2-8. To provide an accurate measurement, a liquid measuring cup must be placed on a level surface and viewed at eye level.

Ladles. A *ladle* is a fixed-size cup attached to a long handle. Ladles come in a variety of sizes and are used for serving controlled portions of liquids such as soups and sauces. The volume (in fluid ounces) held in the cup of the ladle is stamped on the handle. Although ladles can be used as volume measurement tools for ingredients, they are most commonly used to serve food or beverages.

Portion-Controlled Scoops. A *portion-controlled scoop* is a volume measurement tool that consists of a handle with a fixed-size scoop at the end. Portion-controlled scoops are also referred to as "ice cream scoops" or "dishers." Portion-controlled scoops come in a variety of sizes and are marked with a number. The number indicates how many scoops are required to make 1 quart (32 fl oz). **See Figure 2-9.** For example, a #32 scoop would take 32 level-filled scoops to equal 32 fluid ounces at 1 ounce per scoop. The lower the number on the scoop, the fewer scoops it takes to empty a container. For example, a #12 scoop holds 2⅓ fluid ounces and a #6 scoop holds 5⅕ fluid ounces.

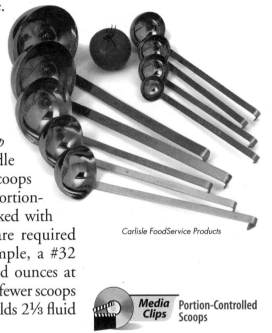

Carlisle FoodService Products

Media Clips Portion-Controlled Scoops

Portion-Controlled Scoops

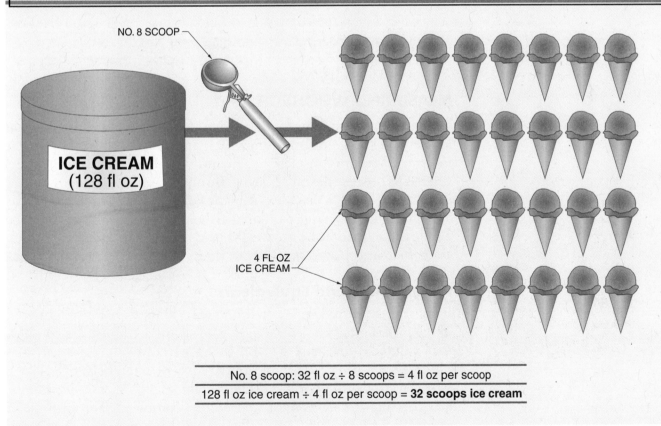

NO. 8 SCOOP

ICE CREAM
(128 fl oz)

4 FL OZ
ICE CREAM

No. 8 scoop: 32 fl oz ÷ 8 scoops = 4 fl oz per scoop

128 fl oz ice cream ÷ 4 fl oz per scoop = **32 scoops ice cream**

Figure 2-9. The number on a portion-controlled scoop indicates how many scoops can be generated from a given volume of food.

1. Define measurement equivalent.

   _____

   _____

2. List the six customary measurement equivalents for volume that every foodservice employee should memorize.

   _____

   _____

3. Describe the proper method for measuring flour using a dry measuring cup.

   _____

   _____

4. Which volume measurement tool is best for measuring liquid ingredients and why?

   _____

   _____

5. How many fluid ounces does a #16 portion-controlled scoop hold?

   _____

   _____

MEASURING WEIGHT

The most common units for measuring weight used in the professional kitchen are the gram (g), ounce (oz), and pound (lb or #). Weight measurements are considered to be more accurate than volume measurements because the individual performing the measurement cannot impact the result. When measuring by volume, a measurement can be affected by the technique used to fill the measuring tool. However, since a scale can only indicate the weight placed on it, the manner in which an ingredient is placed on the scale does not affect the measurement.

Weight Units and Equivalents

Because there are only three common units of measure for weight used in the professional kitchen, weight equivalents are much less complex than volume equivalents. The only customary weight equivalent that must be memorized is that 16 ounces equals 1 pound (16 oz = 1 lb).

When standardized recipes list ingredient weights in customary units, it is common to see pounds and ounces combined. For example, a recipe may call for 1 lb 2 oz of flour. A measurement written this way is to be read "1 pound plus 2 ounces of flour."

The same metric prefixes used in volume measurements apply to weight measurements. For example, 1000 grams equals 1 kilogram just as 1000 liters equals 1 kiloliter. Grams and kilograms are the most common metric unit measurements of weight used in the professional kitchen.

Calculating Weight Equivalents

The only basic customary weight equivalent used in the professional kitchen is 16 ounces = 1 pound. To change a measurement given in pounds (3 lb) to ounces, the number of pounds is multiplied by 16.

3 lb = **48 oz** ($3 \times 16 = 48$)

To change a measurement given in ounces (64 oz) to pounds, the number of ounces is divided by 16.

64 oz = **4 lb** ($64 \div 16 = 4$)

When an ingredient measurement is provided in pounds and ounces, such as 3 lb 2 oz, and the measurement must be changed to ounces only, the first step is to multiply the number of pounds in the original measurement by 16 ($3 \times 16 = 48$). Then, that number is added to the amount of ounces in the original measurement ($48 + 2 = 50$). Therefore, in this example, 3 lb 2 oz = 50 oz.

To change an ingredient measurement from ounces only to pounds and ounces, the number of ounces is divided by 16. Then, the quotient is set equal to the number of pounds and any remainder is equal to the number of ounces. For example, 50 oz $\div$ 16 = 3 with a remainder of 2. This result would be written as 3 pounds 2 ounces.

Sometimes it is necessary to determine equivalent measurements between customary and metric units such as when calculating how many grams there are in 2 pounds. Calculations of this type are discussed in Chapter 4: Converting Measurements and Scaling Recipes.

Tools for Measuring Weight

The tools used for measuring weight are called scales. The three types of scales used in the professional kitchen are mechanical scales, digital scales, and balance scales. **See Figure 2-10.** One thing all scales have in common is that they must be set to zero after a container is placed on them and prior to adding any ingredients into the container. The reason for this is that the weight of the container must be taken into account before accurate measurements can be made. The procedure for setting a scale to zero depends on the type of scale being used.

Equivalent Guide

To change a measurement in kilograms to grams, multiply the number of kilograms by 1000 (1 kg = 1000 g).

2 kg = 2000 g

(2 × 1000 = 2000)

To change a measurement in grams to kilograms, divide the number of grams by 1000 (1000 g = 1 kg).

3000 g = 3 kg

(3000 ÷ 1000 = 3)

Equivalent Guide

The procedure for changing a measurement in ounces to pounds and ounces is as follows:

Step 1: Divide ounces by 16 (16 oz = 1 lb).

```
            QUOTIENT
         3
   16)50
      −48      REMAINDER
        2
```

Step 2: Set the quotient as the number of pounds and the remainder as the number of ounces.

Answer = 3 lb 2 oz

Media Clips   Measuring Weight

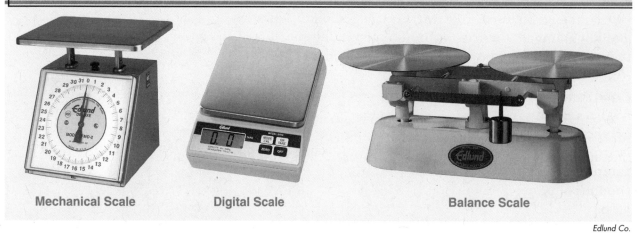

Edlund Co.

Figure 2-10. The three types of scales used in the professional kitchen are mechanical, digital, and balance scales.

Mechanical Scales. A *mechanical scale* is a scale with a spring-loaded platform and a mechanical-dial display. Mechanical scales are also referred to as spring scales or portion scales. Mechanical scales are used for weighing out consistent portions of food as well as measuring ingredients to be used in a recipe.

The dial indicates the measurement capacity (or range) of the scale and the increments in which items can be weighed. Mechanical scales are available that measure in grams, ounces, or pounds. For example, a 2-pound (32 oz) scale can measure ingredients that weigh up to 2 pounds in ¼-ounce increments. The dial will be marked with numbers ranging from 0 ounces to 32 ounces. Marks between each number show the ¼-ounce increments. **See Figure 2-11.**

To measure ingredients on a mechanical scale, an empty container is first placed on the scale platform and the dial is turned so that the needle lines up with the zero. Ingredients are then added to the container. The needle will rotate up the dial as more ingredients are added. Ingredients are no longer added once the desired measurement amount has been indicated on the dial.

Scale Capacity and Measurement Increments

3¼ oz Cheese — PLATFORM

OUNCES

¼ OUNCE INCREMENT

32 oz (¼ oz)

MEASUREMENT INCREMENTS

MAXIMUM CAPACITY

Figure 2-11. A mechanical scale contains a dial that indicates the maximum weight capacity of the scale and the increments in which items can be weighed.

Before a mechanical scale is used, it must be verified that the amount of weight to be measured is less than the maximum capacity of the scale. For example, a 2-pound scale can only be used to measure 2 pounds or less of an ingredient at a time. The smallest mark on the dial for a 2-pound scale is ¼ of an ounce. So, if a recipe calls for an ingredient to be weighed in an increment that is less than ¼ of an ounce, the 2-pound scale will not work. Instead, a scale with smaller increments and a higher capacity would be required to get a more accurate reading.

If the required amount of an ingredient to be weighed is larger than the capacity of the scale, the ingredient must be weighed in parts. For example, if the highest capacity of a scale in a kitchen is 2 pounds, yet 4 pounds of ground beef are to be weighed, 2 pounds of ground beef can be placed on the scale and then transferred to another container. Then another 2 pounds of ground beef are weighed and added to the first 2 pounds already weighed.

Edlund Co.

Digital Scales. A *digital scale* is an electronic scale with a sensor that measures weight and displays the result electronically. Digital scales measure weight in grams, ounces, or pounds. Most digital scales have a switch that is used to select whether the weight is displayed in customary or metric units. They are used for the same purposes as mechanical scales but are especially ideal for use at salad bars and delicatessens where items are sold by weight.

The procedure for using a digital scale is similar to that of a mechanical scale. The only difference is that a digital scale is set to zero by pressing a button instead of turning a dial. Some will have a button labeled "tare." Tare is another term for setting a scale to zero. Digital scales also have increments and maximum capacities but are more accurate than mechanical scales, especially when measuring small increments. The capacity and increments of a digital scale are printed on the faceplate so that the proper scale can be selected for the ingredient to be weighed.

Balance Scales. A *balance scale* is a scale with two platforms that uses a counterbalance system to measure weight. The ingredient to be weighed is placed on one platform and a counterbalance weight is placed on the other platform. The beam between the two platforms has a smaller counterbalance weight that is used for fine adjustments of the scale. The scale is balanced when the weight on the two platforms is equal. A balance scale is also referred to as a baker's scale and is most commonly used in bakeries.

To use a balance scale, an empty container is placed on one platform of the scale. The scale can then be set to zero by adding counterbalance weights to the other platform until the scale is balanced. An additional counterbalance weight is then added to the second platform equal to the desired measurement.

For example, to measure 5 pounds of flour, a 5-pound counterbalance weight is placed on the balance platform opposite the empty container. Flour is then added to the container on the first platform until the scale balances again indicating that 5 pounds of flour has been added to the empty container. The smallest amount that can be weighed using a balance scale is based on the increments indicated on the beam between the two platforms. Technically, a balance scale does not have a maximum operating capacity. Instead it is limited by the amount of counterbalance weights that are provided with the scale. **See Figure 2-12.**

Balance Scales

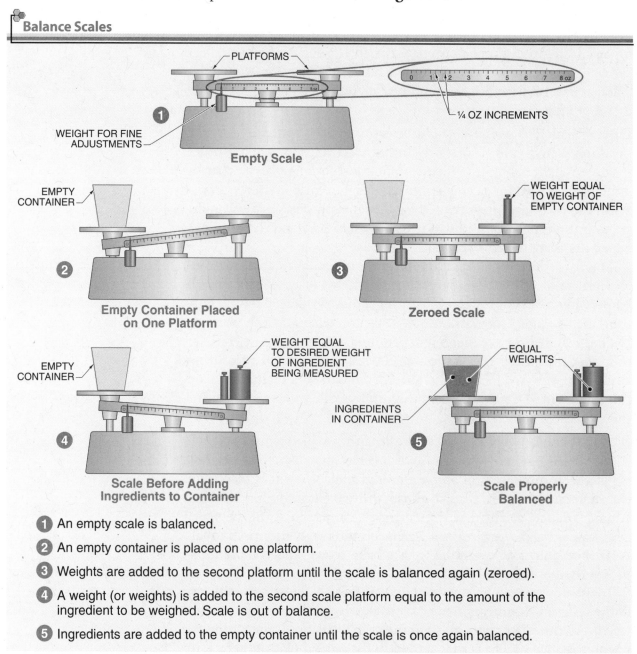

1 An empty scale is balanced.

2 An empty container is placed on one platform.

3 Weights are added to the second platform until the scale is balanced again (zeroed).

4 A weight (or weights) is added to the second scale platform equal to the amount of the ingredient to be weighed. Scale is out of balance.

5 Ingredients are added to the empty container until the scale is once again balanced.

Figure 2-12. A balance (baker's) scale uses a counterbalance system to measure weight.

1. What is the customary weight measurement equivalent that every foodservice employee should know?

_____

2. Why is it important to set a scale to zero before adding ingredients to an empty container placed on the scale?

_____

_____

3. What type of scale displays weight measurements electronically?

_____

4. If 1¼ ounces of baking powder need to be weighed, which scale will give a more accurate result: a mechanical scale with a capacity of 10 pounds and increments of 1 ounce, or a mechanical scale with a capacity of 4 ounces and ⅛-ounce increments?

_____

5. How would 10 pounds of grated cheese be measured using a digital scale that has a capacity of 4 pounds?

_____

_____

_____

6. Change 4 pounds 3 ounces to a measurement in ounces only.

_____

_____

MEASURING TIME AND TEMPERATURE

It is just as important to accurately measure time as it is to measure ingredients. Many recipes provide time measurements in the instructions. Time is measured in days, hours, minutes, or seconds using clocks and timers. For example, recipe instructions may specify that a cake is to bake for 35 minutes or that egg whites are to be beaten for 2 minutes.

Kitchen timers are used in the professional kitchen to sound an alarm when a given amount of time has elapsed. Timers allow a foodservice worker to focus on other tasks without the need to constantly check a clock. Kitchen timers come in a variety of styles including large hanging timers, countertop timers, and digital pocket timers. **See Figure 2-13.**

Hanging Timer

Countertop Timer

Digital Timer

Browne-Halco (NJ)

Figure 2-13. A kitchen timer can be set to sound an alarm when a certain amount of time has passed.

Temperature is another measurement that can lead to the success or failure of a food product. Temperature is measured in the professional kitchen using a thermometer in either the customary unit of degrees Fahrenheit (°F) or the metric unit of degrees Celsius (°C). Common uses of a thermometer include the following:

- checking the internal temperature of food, such as meat, to ensure it has been cooked to a high enough temperature to be served safely
- monitoring the temperature inside an oven
- monitoring the temperature of oil in a deep fryer
- monitoring the temperature in a freezer or refrigerator
- checking the temperature of foods upon receipt to ensure that the food was stored at the proper temperature during transport
- checking the temperature of food held in a steam table or refrigerated table to ensure it is held at the proper temperature and is safe to serve

Temperature can be measured using different styles of thermometers such as instant-read thermometers, candy/deep fry thermometers, electric probe thermometers, or infrared thermometers. **See Figure 2-14.** Instant-read thermometers are the most common type of thermometer used in a professional kitchen. Most cooks keep an instant-read thermometer in their jacket pocket and use it frequently to quickly check food temperatures.

Kitchen Thermometers

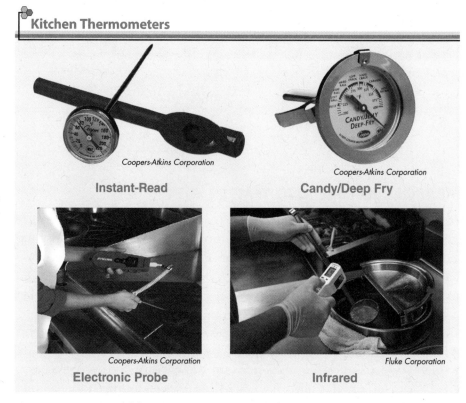

Coopers-Atkins Corporation
Instant-Read

Coopers-Atkins Corporation
Candy/Deep Fry

Coopers-Atkins Corporation
Electronic Probe

Fluke Corporation
Infrared

Figure 2-14. A variety of thermometers are used in the professional kitchen.

Candy/deep fry thermometers are used for measuring higher temperatures and can be clipped to the side of a pot so that the cook can monitor the temperature during the cooking process. Electronic probe thermometers have a temperature probe connected to a digital readout and can be used to measure a wide range of temperatures. Infrared thermometers have sensors to measure the surface temperature of food items without actually coming in contact with the food. They are also recommended for applications such as checking the temperature of cold food held in a salad bar or hot food held in a steam table.

Checkpoint 2-4

1. Explain how a kitchen timer can help a cook to work more efficiently.

2. Explain why a cook would use a thermometer to check the internal temperature of a pork roast.

MEASURING DISTANCE

While distance measurements are not as common in food service as weight and volume measurements, a working knowledge of distance units of measure and equivalents is required. For example, distance measurements are commonly used when determining room sizes and furniture configurations, tabletop set ups, and the sizes of serving platters used at special events. The measurements used for size are the same as those used for distance.

The most commonly used customary units of measure for distance are the inch (in.) and the foot (ft). Feet can also be represented by a prime symbol (') and inches can be represented by a double-prime symbol ("). For example, 2 feet 4 inches is the same as 2'-4".

One important customary measurement equivalent to remember is that 12 inches = 1 foot. The standard metric unit of measure for distance is the meter (m). In food service it is most common to see metric distance measurements in millimeters (mm), centimeters (cm), and meters (m). Just as with volume and weight measurements the metric prefixes indicate that 1 meter = 100 centimeters = 1000 millimeters.

In the professional kitchen, the shape or capacity of cooking equipment is often described by its size (10" sauté pan or a 14" × 20" roasting pan) or by the volume that the equipment can hold (2-quart saucepot or a 10-gallon stockpot). For example, sheet pans and hotel pans are available in standard sizes and are named according to their size.

Hotel pans come in various shapes and depths that provide flexibility for how they are arranged in steam tables, chafing dishes, and refrigerated tables. The largest hotel pan is referred to as a full pan. All other hotel pans are described as some fraction of a full pan. For example, 3 one-third pans occupy the same amount of space in a steam table as one full pan. **See Figure 2-15.** The same consideration applies to sheet pans. A full sheet pan is twice the size of a half sheet pan.

Hotel Pans

Hotel Pan Capacity		
Pan Size	Depth*	Capacity†
Full	2½	8
	4	13
	6	20
⅔	2½	5½
	4	6½
	6	10
½	2½	3½
	4	5½
	6	8
½ long	2½	3½
	4	5½
	6	8
⅓	2½	2½
	4	4
	6	6
¼	2½	2
	4	3
	6	4½
⅙	2½	1
	4	2
	6	2½
⅑	2½	⅝
	4	1⅛

* in inches
† in quarts

The Vollrath Company, LLC

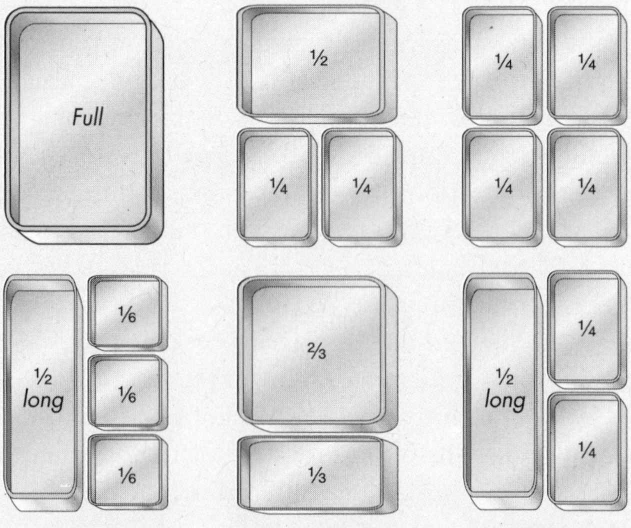

Figure 2-15. The largest hotel pan is a full pan. All other pans are described as a fraction of a full pan.

Checkpoint 2-5

1. If the dimensions of a banquet table are 2 feet by 4 feet, what are the dimensions of the table in inches?

   _____

   _____

2. If a steam table can hold 2 full hotel pans, how many one-sixth pans can the table hold?

   _____

Chapter 2 Summary

One of the main characteristics of a successful foodservice operation is the consistency of the food served. This means that a customer can be assured that a particular food item will look and taste the same on every visit to a particular foodservice operation. Standardized recipes and accurate measurement of the ingredients in each recipe are essential to producing consistent food and beverage products for customers.

In order to measure ingredients accurately, a thorough understanding of the various units of measure and measurement tools used in the professional kitchen is required. Foodservice employees use both customary and metric units of measure. Both systems use standardized units of measure for volume, weight, time, temperature, and distance. It is critical for the successful production of food and beverages to use proper techniques when measuring volume and weight. Math skills are used to calculate measurement equivalents, select the appropriate measuring tool, and perform the actual measuring process.

Checkpoint Answers

Checkpoint 2-1

1. Customers expect the food from a particular foodservice operation to look and taste the same every time it is ordered.
2. A standardized recipe is a list of ingredients, ingredient amounts, and procedural steps for preparing a specific quantity of a food item.
3. A unit of measure is a fixed quantity that is widely accepted as a standard of measurement.
4. Volume is a measurement of the physical space a substance occupies. Weight is a measurement of the heaviness of a substance regardless of its size.

Checkpoint 2-2

1. A measurement equivalent is the amount of one unit of measure that is equal to another unit of measure.
2. Every foodservice employee memorizes the following customary volume measurement equivalents: 4 qt = 1 gal.; 2 pt = 1 qt; 2 c = 1 pt; 8 fl oz = 1 c; 2 tbsp = 1 fl oz; and 3 tsp = 1 tbsp.
3. To measure flour using a dry measuring cup, spoon the flour into the cup, initially overfilling the cup, and then skim off the excess flour by dragging a spatula or the back of a knife over the edge of the cup.
4. A liquid measuring cup is the best tool for measuring liquid ingredients because there is less potential for spilling the liquid ingredient.
5. 2 fl oz/scoop   (32 fl oz ÷ 16 scoops = 2 fl oz/scoop)

Checkpoint 2-3

1. The customary measurement equivalent for weight is 1 lb = 16 oz.
2. A scale must be set to zero prior to adding ingredients to take into account the weight of the container when measuring its contents.
3. A digital scale displays results electronically.
4. The scale with the 4-ounce capacity and ⅛-ounce increments will give the more accurate result.
5. The cheese should be weighed in batches of 4 pounds, 4 pounds, and 2 pounds.   (4 lb + 4 lb + 2 lb = 10 lb)
6. 67 oz   [(4 lb × 16 oz/lb) + 3 oz = 64 oz + 3 oz = 67 oz]

Checkpoint 2-4

1. Since the cook knows the timer will alarm, other tasks can be done without the cook checking the clock frequently.
2. A cook would use a thermometer to make sure that the pork roast has been cooked to a temperature high enough to be served safely.

Checkpoint 2-5

1. 24 in. by 48 in.   (2 ft × 12 in./ft = 24 in., and 4 ft × 12 in./ft = 48 in.)
2. A steam table that holds 2 full hotel pans can also hold 12 one-sixth pans. One full hotel pan is equal to 6 one-sixth pans.

Calculating Measurements

In food service, measurements are more than just a list of the quantities of ingredients used in a recipe. Measurements also relate to the shape and size of kitchen equipment, the amount of space available in a facility, and the number of servings that can be made from a given amount of food. An understanding of how to perform calculations with measurements is an essential skill for any foodservice professional. The importance of good math skills in the professional kitchen becomes more obvious when calculations involve whole numbers, fractions, and decimals.

Chapter Objectives

1. Add, subtract, multiply, and divide measurements involving whole numbers.
2. Convert between improper fractions and mixed numbers.
3. Add, subtract, multiply, and divide fractions.
4. Add, subtract, multiply, and divide decimals.
5. Convert between fractions and decimals.
6. Calculate the area of a rectangular surface and a circular surface.
7. Calculate the volume of a rectangular solid and a cylinder.
8. Explain how angles are used in food service.
9. Calculate the average (mean), median, and mode from a range of numbers.
10. Generate a data table and a graph.

Key Terms

- measurement
- fraction
- numerator
- denominator
- proper fraction
- improper fraction
- mixed number
- common denominator
- lowest common denominator
- reciprocal
- decimal
- rounding
- area
- average

WHOLE NUMBER MEASUREMENTS

A *measurement* is a number with a corresponding unit of measure. For example, 3 gallons, 2 pounds, and 6 inches are all measurements. Calculations involving only numbers are similar to calculations involving measurements. However, when performing basic math operations with measurements, the unit of measure must be included.

Adding and Subtracting Measurements

Addition and subtraction of measurements are straightforward if the unit of measure is the same for every measurement in a calculation. For example, a recipe for punch calls for 3 quarts of cranberry juice, 4 quarts of orange juice, and 2 quarts of club soda. To determine the size of a punch bowl required to hold the entire recipe, the quantities of the individual ingredients must be added together.

$$3 \text{ qt} + 4 \text{ qt} + 2 \text{ qt} = 9 \text{ qt}$$

Therefore, a punch bowl that will hold at least 9 quarts of liquid is required. If, however, a punch recipe uses different volume units of measure, such as 3 quarts of cranberry juice, 1 gallon of orange juice, and 4 pints of club soda, the measurements must be changed to the same unit of measure before they can be added. **See Figure 3-1.**

Carlisle FoodService Products

Media Clips Adding Measurements

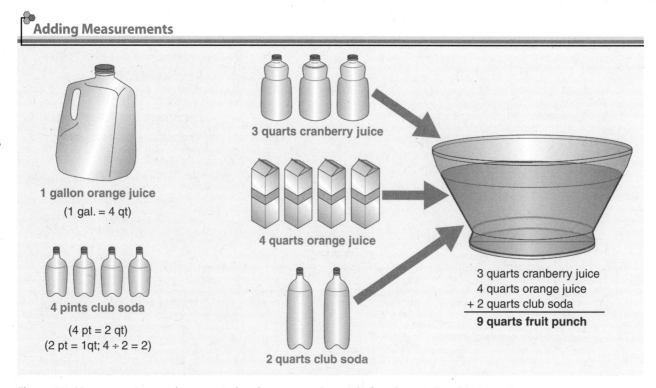

Adding Measurements

1 gallon orange juice
(1 gal. = 4 qt)

4 pints club soda
(4 pt = 2 qt)
(2 pt = 1qt; 4 ÷ 2 = 2)

3 quarts cranberry juice

4 quarts orange juice

2 quarts club soda

3 quarts cranberry juice
4 quarts orange juice
+ 2 quarts club soda
_____
9 quarts fruit punch

Figure 3-1. Measurements must be converted to share a common unit before they can be added together.

If 9 quarts of punch is prepared and 1 pint of punch is left over after a bridal shower, how much punch was consumed? To calculate the amount of punch consumed, 1 pint must be subtracted from 9 quarts. First, 9 quarts is changed to pints (2 pints = 1 quart).

9 qt = 18 pt (9 × 2 = 18)

Then, the 1 pint of punch that was left over is subtracted from the 9 quarts prepared, which results in 17 pints consumed.

18 pt – 1 pt = **17 pt**

Carlisle FoodService Products

Multiplying and Dividing Measurements

In food service, multiplication and division are used when increasing or decreasing a measurement and changing a measurement from one unit to another. When a measurement is multiplied or divided by a number, the unit of measure stays the same. For example, 5 gallons multiplied by 2 is 10 gallons. Likewise, 10 gallons divided by 2 is 5 gallons.

5 gal. × 2 = 10 gal.
and
10 gal. ÷ 2 = 5 gal.

Checkpoint 3-1

1. Define measurement.

   _____

2. What is the sum of 6 cups and 4 pints? Provide the answer in cups.

   _____

   _____

3. Subtract 2 pints from 2 quarts. Provide the answer in quarts.

   _____

   _____

4. Multiply 4 cups by 6.

   _____

   _____

FRACTION MEASUREMENTS

Many measurements in the professional kitchen involve the use of fractions. A *fraction* is a part of a whole. A fraction is written as two numbers (a numerator and a denominator) separated by a line (fraction bar). A *numerator* is the number in a fraction at the top (or to the left) of a fraction bar that represents the parts of a whole. A *denominator* is the number in a fraction at the bottom (or to the right) of a fraction bar that represents the number of parts into which a whole is divided. For example, a quarter-cup dry measuring cup is represented by the fraction ¼ where the 1 is the numerator and 4 is the denominator. Since a fraction represents part of a whole, the fraction ¼ represents one part out of four, ²⁄₄ represents two parts out of four, ¾ represents three parts out of four, and ⁴⁄₄ represents four parts out of four. **See Figure 3-2.**

Media Clips Fractions

Fractions are classified as either proper or improper. A *proper fraction* is a fraction in which the numerator is smaller than the denominator. An *improper fraction* is a fraction in which the numerator is larger than the denominator. For example, ¾ is a proper fraction and ⁴⁄₃ is an improper fraction.

When the numerator and denominator are equal, such as ³⁄₃ or ⁴⁄₄, the value of the fraction is 1. Any whole number can be written as a fraction by using the whole number as the numerator and the denominator equal to 1. For example, the whole number 4 written as a fraction is ⁴⁄₁.

A *mixed number* is a combination of a whole number and a fraction. For example, 1½ cups of milk is 1 cup of milk plus ½ cup of milk. Improper fractions and mixed numbers are related to each other. For example, the improper fraction ⁷⁄₃ can be converted to the mixed number 2⅓ by dividing the numerator by the denominator. Likewise, the mixed number 2⅓ can be converted to the improper fraction ⁷⁄₃.

Fractions (Parts of a Whole)

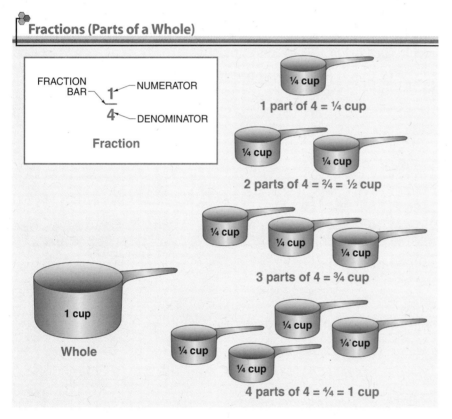

Figure 3-2. A fraction represents part of a whole. The numerator represents the part and the denominator represents the whole.

To convert a mixed number to an improper fraction, the first step is to multiply the whole number by the denominator. For example, in the mixed number 2⅓, the whole number is 2 and the denominator is 3, and the product of those two numbers is 6 (2 × 3 = 6). The next step is to add the original numerator (1) to that product to calculate the numerator of the improper fraction (6 + 1 = 7). Finally, the calculated numerator (7) is placed over the original denominator (3) to form the improper fraction ⅞. **See Figure 3-3.**

Fractions where both the numerator and denominator can be divided evenly by the same number may be reduced to an equivalent fraction. This process is called reducing a fraction to its lowest terms. For example, the fraction ⅝ can be reduced to ¾ by dividing the numerator and the denominator by 2. In the professional kitchen, fractions should be reduced to the lowest terms possible to make measuring easier.

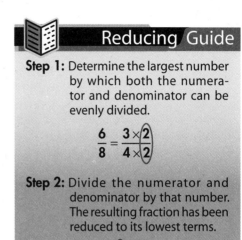

Reducing Guide

Step 1: Determine the largest number by which both the numerator and denominator can be evenly divided.

$$\frac{6}{8} = \frac{3 \times 2}{4 \times 2}$$

Step 2: Divide the numerator and denominator by that number. The resulting fraction has been reduced to its lowest terms.

$$\frac{6 \div 2}{8 \div 2} = \frac{3}{4}$$

Converting Between Improper Fractions and Mixed Numbers

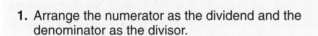

Example: Convert ⅞ to a mixed number.

1. Arrange the numerator as the dividend and the denominator as the divisor.

2. Divide the numerator by the denominator.

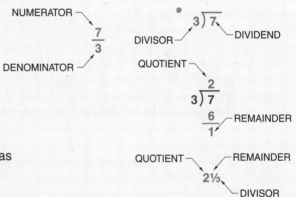

3. Write the mixed number using the quotient as the whole number and the remainder over the divisor as the fraction.

Answer: 2⅓

Example: Convert 2⅓ to an improper fraction.

1. Multiply the whole number by the denominator.

2. Add the product to the numerator and place the sum over the original numerator.

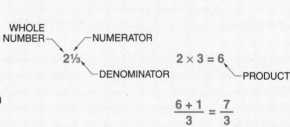

Answer: ⅞

Figure 3-3. Improper fractions can be converted into mixed numbers and mixed numbers can be converted into improper fractions.

Adding and Subtracting Fractions

Adding and subtracting fractions that have a common denominator is simple. A *common denominator* is a denominator that is the same number in two or more fractions. A common denominator must be determined before fractions with unlike denominators can be added or subtracted.

Common Denominators. To add two fractions or more that have a common denominator, the numerators are added together and the sum is placed over the common denominator in the result. For example, if a fruit salad recipe calls for ¼ cup of strawberries, ¼ cup of blueberries, and ¼ cup of raspberries, the total amount of berries required is ¾ cup. **See Figure 3-4.**

Likewise, two fractions that have a common denominator are subtracted by simply subtracting one numerator from the other numerator. The difference is placed over the common denominator. For example, if a fish fillet weighs ⅝ pound and ⅛ pound of skin is removed, the final weight of the fish fillet is ½ pound.

$$\frac{5}{8} - \frac{1}{8} = \frac{4}{8} = \frac{1}{2}$$

Adding and Subtracting Fractions with Like Denominators

Adding

Example: A fruit salad recipe calls for ¼ cup of strawberries, ¼ cup of blueberries, and ¼ cup of raspberries. What is the total amount of berries required?

$$\frac{1}{4} + \frac{1}{4} + \frac{1}{4} = \frac{1+1+1}{4} = \frac{3}{4}$$

SUM OF NUMERATORS

DENOMINATOR STAYS THE SAME

Answer: ¾ cup

Subtracting

Example: A fish fillet weighing ⅝ pound has ⅛ pound of skin removed. What is the final weight of the fish fillet?

$$\frac{5}{8} - \frac{1}{8} = \frac{5-1}{8} = \frac{4}{8} = \frac{4 \div 4}{8 \div 4} = \frac{1}{2}$$

DIFFERENCE BETWEEN NUMERATORS

DENOMINATOR STAYS THE SAME

Answer: ½ pound

Figure 3-4. When all the fractions in a calculation involving addition or subtraction have a common denominator, the numerators are added or subtracted and the denominator stays the same.

Unlike Denominators. In order to add or subtract fractions with unlike denominators, the fractions must be rewritten so that they have a common denominator. The process of rewriting the fractions to be added or subtracted begins by finding the lowest common denominator. The *lowest common denominator* is the smallest number into which each of the denominators of a group of fractions can divide evenly.

For example, if a spice mixture recipe requires ½ cup of black pepper, ⅔ cup of garlic powder, and ¾ cup of salt, how many total cups of the spice mixture will the recipe make? In order to add the three fractions together, the lowest common denominator among the three denominators (2, 3, and 4) must be found first. Since 12 is the smallest number that 2, 3, and 4 all divide into evenly, 12 is the lowest common denominator. **See Figure 3-5.**

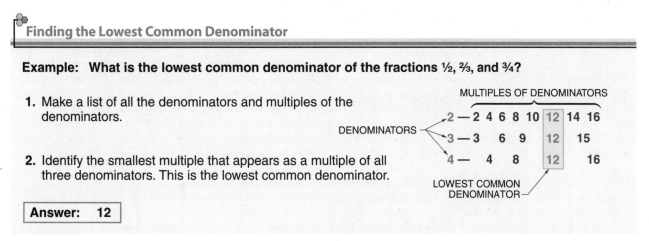

Finding the Lowest Common Denominator

Example: What is the lowest common denominator of the fractions ½, ⅔, and ¾?

1. Make a list of all the denominators and multiples of the denominators.

2. Identify the smallest multiple that appears as a multiple of all three denominators. This is the lowest common denominator.

Answer: 12

Figure 3-5. Before fractions with unlike denominators can be added or subtracted, a lowest common denominator must be determined.

The next step is to rewrite all the fractions so that they include the lowest common denominator. To rewrite each fraction as a fraction of equal value with a common denominator of 12, the numerator and denominator must be multiplied by the same number. For example, since 2 × 6 = 12, ½ is multiplied by 6/6.

$$\frac{1}{2} \times \frac{6}{6} = \frac{1 \times 6}{2 \times 6} = \frac{6}{12}$$

Then, ⅔ is multiplied by 4/4 to get 8/12 (⅔ × 4/4 = 8/12). Finally, ¾ is multiplied by 3/3 to get 9/12 (¾ × 3/3 = 9/12). Now that the fractions 6/12, 8/12, and 9/12 have a common denominator, they can be added by adding the numerators (6 + 8 + 9 = 23) and placing the sum (23) over the common denominator (12). Therefore, the total amount of spice mix the recipe makes is 23/12 cups. Then, 23/12 cups can be converted to a mixed number by dividing the numerator by the denominator (23 ÷ 12 = 1 with a remainder of 11 or 1 11/12 cups). **See Figure 3-6.**

Example: A spice mixture recipe requires ½ cup of black pepper, ⅔ cup of garlic powder, and ¾ cup of salt. How many total cups of the spice mixture will the recipe make?

$$\frac{1}{2} + \frac{2}{3} + \frac{3}{4} =$$ — UNLIKE DENOMINATORS

1. Find the lowest common denominator (12).

$$12 \div 2 = 6$$
$$12 \div 3 = 4$$
$$12 \div 4 = 3$$ — QUOTIENTS

2. Divide the lowest common denominator by each of the unlike denominators.

3. Multiply the numerator and denominator of each fraction by the corresponding quotients from Step 2 to rewrite the fractions as equivalent fractions with a common denominator.

$$\frac{1}{2} \times \frac{6}{6} = \frac{1 \times 6}{2 \times 6} = \frac{6}{12}$$

$$\frac{2}{3} \times \frac{4}{4} = \frac{2 \times 4}{3 \times 4} = \frac{8}{12}$$

$$\frac{3}{4} \times \frac{3}{3} = \frac{3 \times 3}{4 \times 3} = \frac{9}{12}$$

COMMON DENOMINATOR

4. Add the numerators of the equivalent fractions and record the sum over the common denominator.

$$\frac{6 + 8 + 9}{12} = \frac{23}{12}$$

5. Convert answer to a mixed number if necessary.

$$12\overline{)23} = 1\frac{11}{12}$$
$$\underline{-12}$$
$$11$$

Answer: ²³⁄₁₂ cups, or 1¹¹⁄₁₂ cups

Figure 3-6. Fractions with unlike denominators must be rewritten as fractions of equivalent value with common denominators before they can be added together.

Finally, if ½ cup of the spice mix is set aside for use during lunch service and only ⅓ cup is used, how much spice mix will be left? To calculate this amount, ⅓ cup must be subtracted from ½ cup. As with addition, subtraction cannot be performed until a common denominator is found between ½ and ⅓. The same steps are followed as for addition except the numerators of the rewritten fractions are subtracted instead of added. **See Figure 3-7.**

Multiplying Fractions

When multiplying fractions, the numerators and the denominators are multiplied separately. Any mixed number must be converted to an improper fraction before multiplying. The resulting fraction can be converted back to a mixed number after multiplying, if necessary.

For example, if a recipe calls for 2⅓ cups of flour and 1½ times the recipe needs to be prepared, how much flour is needed? To multiply the mixed numbers 2⅓ and 1½, first convert each to the improper fractions ⅞ and ³⁄₂. Next, the numerators are multiplied by each other and the denominators are multiplied by each other. The resulting fraction (²¹⁄₆) can then be converted to a mixed number (3½). Therefore, 3½ cups of flour is needed for the recipe. **See Figure 3-8.**

Example: If ½ cup of a spice mixture is prepared and ⅓ cup of the mixture is used during lunch service, how much spice mixture remains?

$$\frac{1}{2} - \frac{1}{3} =$$ — UNLIKE DENOMINATORS

1. Find the lowest common denominator (6).

2. Divide the lowest common denominator by each of the unlike denominators.

$$6 \div \boxed{2} = 3$$
$$6 \div \boxed{3} = 2$$ — QUOTIENTS

3. Multiply the numerator and denominator of each fraction by the corresponding quotients from Step 2 to rewrite the fractions as equivalent fractions with a common denominator.

$$\frac{1}{2} \times \frac{3}{3} = \frac{1 \times 3}{2 \times 3} = \frac{3}{6}$$
$$\frac{1}{3} \times \frac{2}{2} = \frac{1 \times 2}{3 \times 2} = \frac{2}{6}$$ — COMMON DENOMINATOR

4. Subtract the numerators of the equivalent fractions and record the difference over the common denominator.

$$\frac{3-2}{6} = \frac{1}{6}$$

Answer: ⅙ cup

Figure 3-7. Fractions with unlike denominators must be rewritten as fractions of equivalent value with common denominators before they can be subtracted.

◊◊ **Multiplying Fractions**

Example: A recipe calls for 2⅓ cups of flour and 1½ times the recipe needs to be made. How much flour will be needed?

$$2\frac{1}{3} \times 1\frac{1}{2} =$$

1. Convert both mixed numbers to improper fractions.

$$2\frac{1}{3} = \frac{(2 \times 3) + 1}{3} = \frac{6+1}{3} = \frac{7}{3}$$
$$1\frac{1}{2} = \frac{(1 \times 2) + 1}{2} = \frac{2+1}{2} = \frac{3}{2}$$

2. Multiply the numerators by each other and the denominators by each other.

$$\frac{7}{3} \times \frac{3}{2} = \frac{7 \times 3}{3 \times 2} = \frac{21}{6}$$ — PRODUCT

3. Convert product to a mixed number and reduce to lowest terms.

$$\frac{21}{6} = 6\overline{)21} = 3\frac{3}{6} = 3\frac{1}{2}$$
$$\underline{18}$$
$$3$$

Answer: 3½ cups

Figure 3-8. When multiplying fractions, the numerators are multiplied by each other and the denominators are multiplied by each other.

Dividing Fractions

Dividing one fraction by another fraction involves using a reciprocal. A *reciprocal* is a fraction that is the result of switching the places of the numerator and denominator in a fraction. For example, the reciprocal of ⅔ is ³⁄₂. The product of a fraction and its reciprocal is always 1.

$$\frac{2}{3} \times \frac{3}{2} = \frac{6}{6} = 1$$

To divide fractions, multiply the first fraction (the dividend) by the reciprocal of the second fraction (the divisor). For example, to divide ½ by ⅗, simply multiply ½ by the reciprocal of ⅗ (which is ⅗). The result is ⅚.

$$\frac{1}{2} \div \frac{3}{5} = \frac{1}{2} \times \frac{5}{3} = \frac{5}{6}$$

Likewise, in order to calculate how many ¼-pound hamburger patties a butcher can make out of 12½ pounds of ground beef, 12½ is divided by ¼. **See Figure 3-9.** The first step in dividing the fractions is to convert the mixed number 12½ to an improper fraction, which is ²⁵⁄₂. Next, ²⁵⁄₂ can be divided by ¼. This is done by multiplying ²⁵⁄₂ by ⁴⁄₁ (the reciprocal of ¼) to obtain the result ¹⁰⁰⁄₂. In this case, the result is equal to the whole number 50. Therefore, the butcher can make 50 hamburger patties at ¼ pound each out of 12 ½ pounds of ground beef.

Carlisle FoodService Products

Dividing Fractions

Example: How many ¼-pound hamburger patties can be made from 12½ pounds of ground beef?

$$12\frac{1}{2} \div \frac{1}{4} =$$
DIVIDEND — DIVISOR

1. Convert mixed number to an improper fraction.

$$12\frac{1}{2} = \frac{(12 \times 2) + 1}{2} = \frac{24 + 1}{2} = \frac{25}{2}$$

2. Multiply the dividend by the reciprocal of the divisor.

$$\frac{25}{2} \times \boxed{\frac{4}{1}} = \frac{25 \times 4}{2 \times 1} = \frac{100}{2}$$

RECIPROCAL OF DIVISOR — PRODUCT

3. Reduce the product to lowest terms.

$$\frac{100}{2} = 50$$

Answer: 50 hamburger patties

Figure 3-9. Dividing fractions requires converting the divisor fraction into its reciprocal.

1. What is a fraction?

_____

_____

2. A fraction contains a numerator and a denominator. Which is on top of the fraction bar and which is under the fraction bar?

_____

_____

3. What is a mixed number?

_____

_____

4. Convert $^{25}/_3$ to a mixed number.

_____

_____

5. Convert $2^5/_8$ to an improper fraction.

_____

_____

6. Reduce $^9/_{12}$ to its lowest possible terms.

_____

_____

7. A recipe calls for $12^1/_2$ cups of oats. If two times the recipe needs to be made, what amount of oats will be required?

_____

_____

8. How many $^1/_3$-pound portions of potato salad can be made from $9^2/_3$ pounds of potato salad?

_____

_____

DECIMAL MEASUREMENTS

Just like calculations that contain fractions, calculations that contain decimals require more math skills. In food service, decimal measurements are involved when using metric units of measure and when working with digital scales. An understanding of decimals is also required when

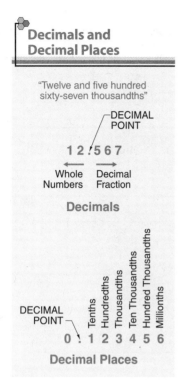

Decimals and Decimal Places

"Twelve and five hundred sixty-seven thousandths"

DECIMAL POINT

1 2 . 5 6 7

Whole Numbers | Decimal Fraction

Decimals

DECIMAL POINT — Tenths, Hundredths, Thousandths, Ten Thousandths, Hundred Thousandths, Millionths

0 . 1 2 3 4 5 6

Decimal Places

Figure 3-10. In a decimal, whole numbers are located to the left of the decimal point and the decimal part is located to the right of the decimal point.

performing calculations involving money. A *decimal* is a number that represents part of a whole and can be expressed as a fraction with a denominator that is a power of 10 (10, 100, 1000, and so on). For example, 0.1 equals ¹⁄₁₀, 0.25 equals ²⁵⁄₁₀₀, and 0.575 equals ⁵⁷⁵⁄₁₀₀₀, which are all parts of the whole number 1.

$$\frac{1}{10} = 0.1 \qquad \frac{25}{100} = 0.25 \qquad \frac{575}{1000} = 0.575$$

The period in the decimal is called a decimal point. The whole number is written to the left of the decimal point and the decimal part is written to the right of the decimal point. **See Figure 3-10.** For example, 2.3 represents the whole number 2 and ³⁄₁₀, 4.35 represents the whole number 4 and ³⁵⁄₁₀₀, and 12.567 represents the whole number 12 and ⁵⁶⁷⁄₁₀₀₀.

$$2\frac{3}{10} = 2.3 \qquad 4\frac{35}{100} = 4.35 \qquad 12\frac{567}{1000} = 12.567$$

When reading or writing a decimal, the decimal point is read and written as "and." The digits to the right of the decimal point have the same place value as the last digit. For example, 12.567 would be read and written as "twelve and five hundred sixty-seven thousandths."

Many monetary values are based on decimals. For example, the dollar ($1.00) is valued at 100 cents. Each cent (penny) is ¹⁄₁₀₀ of a dollar ($0.01) and each nickel is ⁵⁄₁₀₀ of a dollar ($0.05). Likewise, each dime is ¹⁰⁄₁₀₀ or ¹⁄₁₀ of a dollar ($0.10). Each quarter is ²⁵⁄₁₀₀ or ¼ of a dollar ($0.25). Decimals are also used in calculations that involve money such as adding a customer's bill or making change.

Rounding Decimals

Rounding is the process of reducing the number of places in a decimal to achieve a certain degree of accuracy. Decimals can be rounded to any number of places. For example, a decimal rounded to three places (the thousandths place) represents a higher degree of accuracy than a decimal rounded to two places (the hundredths place).

Decimals that end in four or less (4, 3, 2, 1, and 0) are typically rounded down and decimals that end in five or more (5, 6, 7, 8, and 9) are rounded up. For example, the decimal 3.5648 rounded to the hundredths place would be 3.56 since the first digit to the right of the hundredths place is a 4. Therefore, 3.5648 is closer to 3.56 than 3.57. The same decimal rounded to the thousandths place would be 3.565 because the digit in the ten-thousandths place is equal to, or more than, five. **See Figure 3-11.**

In food service, decimals beyond the thousandths place are rarely used. Sometimes calculations involving money are carried out to the thousandths place for additional accuracy. "Mill" is the term used to represent ¹⁄₁₀₀₀ of a dollar ($0.001) or ¹⁄₁₀ of a penny.

Adding and Subtracting Decimals

To add or subtract decimals, the decimal points are first aligned. Then, either addition or subtraction is performed normally as with whole numbers. The decimal point does not move.

For example, three honeydew melons are weighed separately on a digital scale and the results are 6.35 pounds, 5.7 pounds, and 5.24 pounds. The total weight when added together is 17.29 pounds. **See Figure 3-12.** If, after peeling and deseeding the honeydew melons, a total of 13.1 pounds of edible honeydew melon is left, the amount of waste is calculated to be 4.19 pounds by subtracting 13.1 pounds from the original 17.29 pounds.

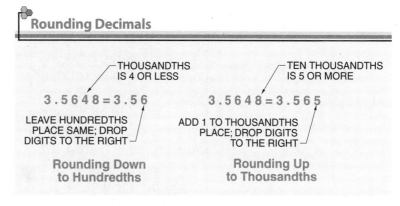

Rounding Decimals

THOUSANDTHS IS 4 OR LESS

3 . 5 6 4 8 = 3 . 5 6

LEAVE HUNDREDTHS PLACE SAME; DROP DIGITS TO THE RIGHT

Rounding Down to Hundredths

TEN THOUSANDTHS IS 5 OR MORE

3 . 5 6 4 8 = 3 . 5 6 5

ADD 1 TO THOUSANDTHS PLACE; DROP DIGITS TO THE RIGHT

Rounding Up to Thousandths

Figure 3-11. Decimals are rounded up or down depending on the value of the last digit in the decimal.

Adding and Subtracting Decimals

Adding

Example: Three honeydew melons weigh 6.35 pounds, 5.7 pounds, and 5.24 pounds. What is the total weight of all three honeydew melons?

$$6.35 + 5.7 + 5.24 =$$

1. Stack numbers and align on decimal points.

```
   6 . 3 5
   5 . 7
 + 5 . 2 4
```

2. Add digits column by column.

```
   ¹
   6 . 3 5
   5 . 7
 + 5 . 2 4
 1 7 . 2 9
```

Answer: 17.29 pounds

Subtracting

Example: After trimming 17.29 pounds of honeydew melon, 13.1 pounds of edible fruit is obtained. How much waste was generated?

$$17.29 - 13.1 =$$

1. Stack numbers and align on decimal points.

```
 1 7 . 2 9
 - 1 3 . 1
```

2. Subtract digits column by column.

```
 1 7 . 2 9
 - 1 3 . 1
   4 . 1 9
```

Answer: 4.19 pounds

Figure 3-12. Decimals are added and subtracted in the same way as whole numbers.

Multiplying Decimals

Decimals are multiplied in the same way as whole numbers. However, placement of the decimal point in the final product is based on the total number of decimal places in the original calculation. For example, to find the total amount of tomato paste in 4.5 cans when the contents of each can weigh 6.5 pounds, 4.5 is multiplied by 6.5. The total is determined by multiplying 6.5 and 4.5 as if they were whole numbers (65 × 45 = 2925). Then, since there are a total of two decimal places (one in 6.5 and one in 4.5), the decimal point is inserted two places from the right. The final answer is 29.25 pounds. **See Figure 3-13.**

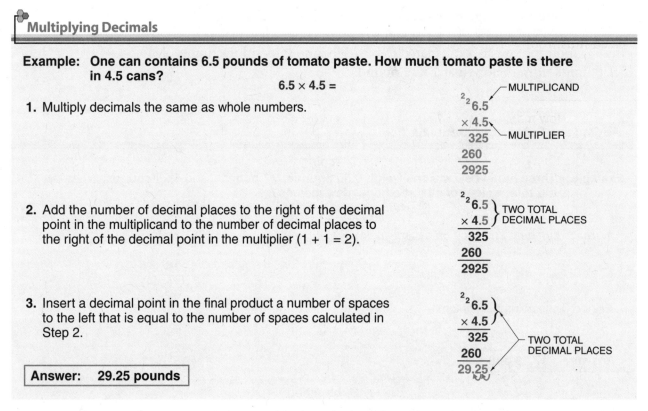

Multiplying Decimals

Example: One can contains 6.5 pounds of tomato paste. How much tomato paste is there in 4.5 cans?

$$6.5 \times 4.5 =$$

1. Multiply decimals the same as whole numbers.

2. Add the number of decimal places to the right of the decimal point in the multiplicand to the number of decimal places to the right of the decimal point in the multiplier (1 + 1 = 2).

3. Insert a decimal point in the final product a number of spaces to the left that is equal to the number of spaces calculated in Step 2.

Answer: 29.25 pounds

Figure 3-13. When multiplying decimals, the location of the decimal point in the product is based on the number of decimal places in the decimals being multiplied.

Dividing Decimals

To divide decimals, the first step is to move the decimal point in the divisor to the right until the divisor becomes a whole number. Next, move the decimal point in the dividend the same number of places to the right. If there are not enough digits in the dividend, add zeroes to the places created by moving the decimal point.

For example, to divide 7.5 by 0.25, the decimal point is moved to the right two places in each number (0.25 becomes 25 and 7.5 becomes 750).

$$0.25\overline{)7.5} = 0.\underset{\curvearrowright\curvearrowright}{25}\overline{)7.\underset{\curvearrowright\curvearrowright}{50}} = 25.\overline{)750}.$$

Then, the decimal point in the dividend is brought up to the quotient and the division is performed the same way as with whole numbers.

$$25\overline{)750}. = 25\overline{)750.} = 2\,5\overline{)\begin{array}{c} 3\,0. \\ 7\,5\,0 \\ \hline 7\,5 \\ \hline 0 \end{array}}$$

Likewise, to determine the number of 0.25-pound (¼-pound) portions that can be obtained from 13.5 pounds of sliced mushrooms, 13.5 is divided by 0.25. Thus, 54 portions can be made from 13.5 pounds of sliced mushrooms. **See Figure 3-14.**

InterMetro Industries Corporation

Dividing Decimals

Example: How many 0.25-pound portions of sliced mushrooms can be obtained from 13.5 pounds of sliced mushrooms?

$$13.5 \div 0.25 =$$

1. Move the decimal point in the divisor to the right enough places to make the divisor a whole number.

   DIVISOR
   $$25.\overline{)13.5}$$

2. Move the decimal point in the dividend the same amount of places to the right.

   DIVIDEND
   $$25.\overline{)1350.}$$

3. Bring the decimal point in the dividend up to the quotient.

   $$25.\overline{)1350.}$$

4. Divide the dividend by the divisor.

   QUOTIENT
   $$25.\overline{)\begin{array}{c} 54.0 \\ 1350. \\ \hline 125 \\ \hline 100 \\ \hline 100 \\ \hline 0 \end{array}}$$

Answer: 54 portions

Figure 3-14. When the divisor and dividend in a division calculation are both multiplied by the same factor of 10, the value of the quotient does not change.

The process of moving the decimal point to the right in both the divisor and the dividend is the same as multiplying both numbers by powers of 10. For example, moving the decimal point one place to the right is the same as multiplying by 10. Moving two places to the right is the same as multiplying by 100. Since both the divisor and the dividend are multiplied by the same power of 10, the value of the quotient is not affected.

Converting Between Fractions and Decimals

To change a fraction to a decimal, divide the numerator by the denominator. For example, if ⅝ of an ounce of salt needs to be measured on a digital scale that displays weight in decimals, the fraction ⅝ can be converted to a decimal by dividing the numerator by the denominator. **See Figure 3-15.** Since the decimal equivalent of ⅝ is calculated to be 0.625, ⅝ of an ounce of salt can be measured as 0.625 ounces on a digital scale.

Converting Fractions to Decimals

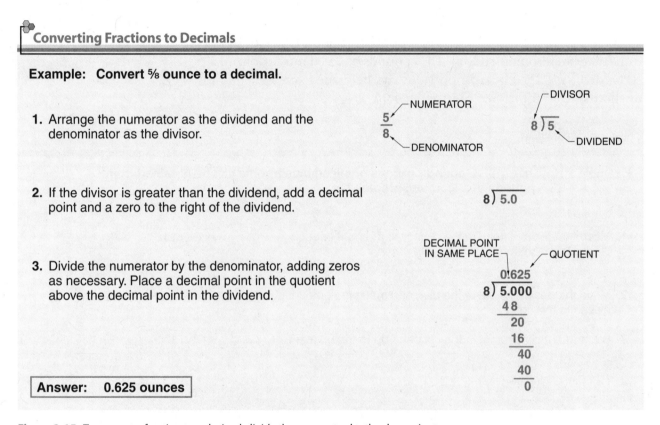

Example: Convert ⅝ ounce to a decimal.

1. Arrange the numerator as the dividend and the denominator as the divisor.

2. If the divisor is greater than the dividend, add a decimal point and a zero to the right of the dividend.

3. Divide the numerator by the denominator, adding zeros as necessary. Place a decimal point in the quotient above the decimal point in the dividend.

Answer: 0.625 ounces

Figure 3-15. To convert a fraction to a decimal, divide the numerator by the denominator.

To change a decimal to a fraction, first the decimal is set as the numerator in a fraction without a decimal point. Then, the denominator is calculated as a power of 10 depending on how many places there are in the decimal being converted. For example, to convert the measurement 0.75 pounds of sugar to a fraction, use 75 as the numerator. Since there are two places in the decimal 0.75, the denominator is calculated as two powers of 10 (10 × 10 = 100). With the denominator set as 100, the fraction becomes ⁷⁵⁄₁₀₀. After reducing the fraction ⁷⁵⁄₁₀₀ to the lowest terms, the final answer is ¾. **See Figure 3-16.**

Converting Decimals to Fractions

Example: Convert 0.75 pounds to a fraction.

1. Set the decimal as the numerator in a fraction (without a decimal point).

2. Set the denominator equal to a power of 10 based on the number of decimal places in the decimal being converted.

3. Reduce the fraction to lowest possible terms. (Convert an improper fraction to a mixed number if necessary.)

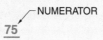

NUMERATOR

$$\underline{75}$$

$$0.75 \atop \text{TWO DECIMAL PLACES}$$

$$\frac{75}{100}$$

DENOMINATOR SET TO TWO POWERS OF 10 (10 × 10)

$$\frac{75}{100} = \frac{25 \times 3}{25 \times 4} = \frac{25}{25} \times \frac{3}{4} = 1 \times \frac{3}{4} = \frac{3}{4}$$

| **Answer:** ¾ **pounds** |

Figure 3-16. Any decimal can be converted to a fraction by setting the denominator as a power of 10.

Checkpoint 3-3

1. Define the term decimal.

2. Define rounding.

3. Which represents a higher degree of accuracy: a decimal rounded to the tenths place or a decimal rounded to the hundredths place?

4. Write out the decimal 12.56 in words.

5. Write down the decimal that represents six and two tenths.

6. Round the decimal 34.6575 to the thousandths place.

7. Three beef tenderloins are to be prepared, one weighs 8.7 pounds, another weighs 7.4 pounds, and the last weighs 8.2 pounds. What is the total weight of the three beef tenderloins?

8. If a pork roast weighed 15.2 pounds before being trimmed and 13.8 pounds afterward, how much waste was trimmed off the roast?

9. If one can of ketchup contains 12.96 cups of ketchup. How many cups of ketchup are there in 3.5 cans?

_____

_____

10. How many 2.5-ounce portions of salad dressing can be created from 62.5 ounces of salad dressing?

_____

_____

11. A recipe calls for 0.375 pounds of sugar. What fraction represents the same amount of sugar?

_____

_____

12. A recipe calls for ⅔ pound of cornstarch. If the cornstarch is weighed on a digital scale, what decimal number would be equal to ⅔ pound? Round the answer to two decimal places.

_____

_____

CALCULATING AREA, VOLUME, AND ANGLES

Measurements can be multiplied to calculate the *area* of a two-dimensional (flat) surface or the *volume* of a three-dimensional object. In food service, two-dimensional surfaces include baking sheets, tabletops, and the floor space of a banquet room for example. Three-dimensional objects include items such as food containers and stockpots as well as walk-in coolers. Angles are used in food service to describe how to cut or portion certain foods and how to properly use some kitchen tools.

Calculating Area

Two distance measurements with the same unit of measure are multiplied by each other to calculate the area of a surface. *Area* is the size of a two-dimensional (flat) surface.

To calculate the area of a square or rectangle, the following formula is applied:

$$A = L \times W$$

where
A = area
L = length
W = width

Area = Length × Width

When calculating area, the numbers in the two measurements are multiplied by each other and the units of measure are multiplied by each other. Therefore, the area unit of measure is the square of the original unit. For example, 2 inches × 2 inches = 4 square inches, 2 feet × 3 feet = 6 square feet, and 3 meters × 3 meters = 9 square meters. Square units of measure are abbreviated two different ways. **See Figure 3-17.**

Calculating Area

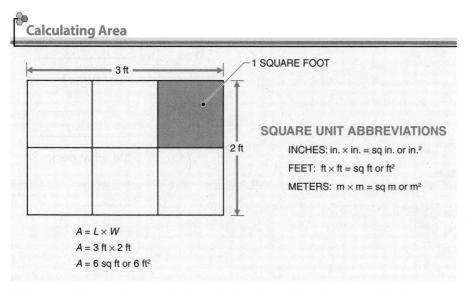

$A = L \times W$
$A = 3 \text{ ft} \times 2 \text{ ft}$
$A = 6 \text{ sq ft or } 6 \text{ ft}^2$

Figure 3-17. The area of a rectangle is equal to length × width and is expressed in square units of measure.

To calculate the area of the floor space of a given banquet room, apply the formula for the area of a rectangle using these problem-solving steps.

Problem-Solving Steps

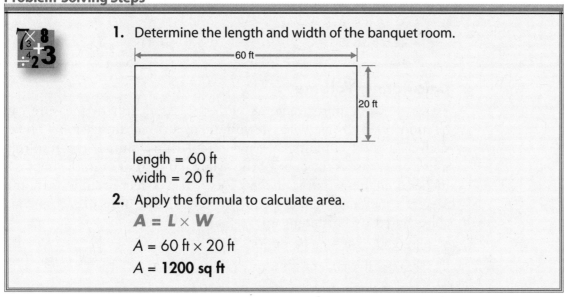

1. Determine the length and width of the banquet room.

length = 60 ft
width = 20 ft

2. Apply the formula to calculate area.

$A = L \times W$

$A = 60 \text{ ft} \times 20 \text{ ft}$

$A = \textbf{1200 sq ft}$

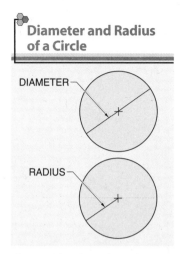

To calculate the area of a circle, the radius of the circle is first multiplied by itself (squared) and then multiplied by the constant number 3.14, shown in formulas as the Greek letter pi (π). The *radius* is the distance from the center of a circle to the edge of the circle and is equal to one half the diameter of the circle. A diameter of a circle is the length of a line drawn through the center of the circle to both edges. **See Figure 3-18.** To calculate the area of a circle, the following formula is applied:

$$A = \pi \times r^2$$

where

A = area

π = 3.14 (constant number)

r = radius

$$Area = 3.14 \times Radius^2$$

Figure 3-18. The radius of a circle is equal to one half the diameter of a circle.

To calculate the area of a given dining table, apply the formula for the area of a circle using these problem-solving steps.

Problem-Solving Steps

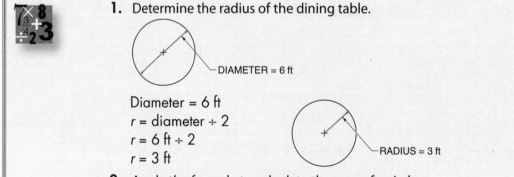

1. Determine the radius of the dining table.

DIAMETER = 6 ft

RADIUS = 3 ft

Diameter = 6 ft

r = diameter ÷ 2

r = 6 ft ÷ 2

r = 3 ft

2. Apply the formula to calculate the area of a circle.

$$A = \pi \times r^2$$

$A = \pi \times (3\ ft)^2 = \pi \times 9$ sq ft

$A = 3.14 \times 9$ sq ft

$A =$ **28.26 sq ft**

Calculating Volume

Three distance measurements with the same unit of measure can be multiplied to calculate the volume of a three-dimensional object such as a food container. When calculating volume, the resulting unit of measure is the cube of the original unit. For example inches × inches × inches = cubic inches, feet × feet × feet = cubic feet, and meters × meters × meters = cubic meters. Cubic units of measure are abbreviated two different ways.

in. × in. × in = cu in. or in^3

ft × ft × ft = cu ft or ft^3

m × m × m = cu m or m^3

In food service, it is common to refer to certain pieces of equipment by their capacity such as a 20-quart stockpot or a 250 cubic foot walk-in cooler. *Capacity* is the volume that can be placed inside an empty three-dimensional object. For example, a steam kettle that has a capacity of 30 quarts can hold a volume of liquid equal to 30 quarts. Likewise, a 20-cubic foot refrigerator can hold ten 2-cubic foot boxes of food.

The volume equivalent between cubic inches and quarts is 57.8 cubic inches = 1 quart. Therefore, to change a volume measured in cubic inches to quarts (a smaller unit to a larger unit), the number of cubic inches is divided by 57.8. For example, to calculate the volume, in quarts, of a stockpot with a volume of 1000 cubic inches, simply divide 1000 by 57.8.

1000 cu in. = **17.3 qt** (1000 ÷ 57.8 = 17.3)

To calculate the volume of a rectangular object, the following formula is applied:

$V = L \times W \times H$

where

V = volume
L = length
W = width
H = height (or depth)

Volume = Length × Width × Height

Carlisle FoodService Products

To calculate the capacity of a given walk-in cooler, apply the formula for calculating the volume of a rectangular object using these problem-solving steps.

Problem-Solving Steps

1. Determine the length, width, and height of the walk-in cooler shown below.

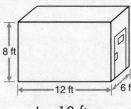

8 ft

12 ft 6 ft

$L = 12$ ft
$W = 6$ ft
$H = 8$ ft

2. Apply the formula for determining the volume of a rectangular object.
$V = L \times W \times H$
$V = 12$ ft $\times 6$ ft $\times 8$ ft
$V =$ **576 cu ft**

To calculate the volume of an object shaped like a cylinder, such as a stockpot, the area of the circular base (or opening) of the object is multiplied by the height (or depth) of the object. The following formula is used to calculate the volume of a cylinder:

$$V = \pi \times r^2 \times H$$

where

V = volume
π = 3.14 (constant number)
r = radius of the base
H = height

Volume = 3.14 × Radius² × Height

To calculate the volume in quarts of a given stockpot, apply the formula for calculating the volume of a cylinder using these problem-solving steps.

Problem-Solving Steps

1. Determine the radius and height of the stockpot shown below.

    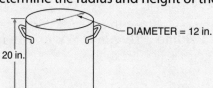

    DIAMETER = 12 in.

    20 in.

    r = diameter ÷ 2
    r = 12 in. ÷ 2 = 6 in.
    H = 20 in.

2. Apply the formula for calculating the volume of a cylinder.

    $$V = \pi \times r^2 \times H$$

    V = 3.14 × (6 in.)² × 20 in.
    V = 3.14 × 6 in. × 6 in. × 20 in.
    V = 3.14 × 720 cu in.
    V = 2260.8 cu in.

3. Change 2260.8 cubic inches to quarts (57.8 cu in. = 1 qt).

    2260.8 ÷ 57.8 = 39.1
    39.1 quarts

Calculating Angles

An *angle* is a measurement that indicates the relationship between two lines when the lines intersect one another. Consider a circle that has been divided into four equal parts by drawing a horizontal line and a vertical line down the center, such as when a pizza is cut into four equal slices. Angles, and more specifically, central angles are created where the lines cross (intersect) at the center of the circle.

Angles are measured in degrees and are abbreviated using the degree symbol (°). A full circle has 360 degrees (360°). When a circle is divided into two equal pieces, each piece of the circle has 180 degrees (360 degrees ÷ 2 = 180 degrees). If one-half of the circle is cut in half again, each of those two pieces would have 90 degrees (180 degrees ÷ 2 = 90 degrees) and central angles would be created that measure 90 degrees each. Regardless of the number of equal pieces a circle is divided into through the center, each piece will have an equal number of degrees and a central angle equal to the same number. **See Figure 3-19.**

Degrees and Angles

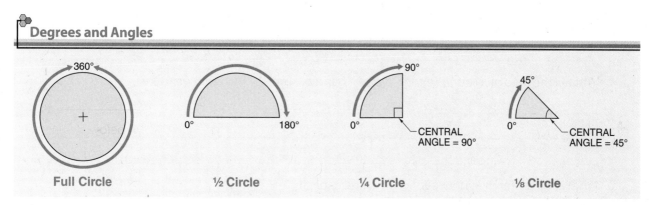

| Full Circle | ½ Circle | ¼ Circle | ⅛ Circle |

Figure 3-19. When a circle is divided into equal portions through the center, central angles are created that equal the number of degrees that the portion of the circle contains.

In addition to specifying how a round shape is cut into equal pieces, angles are also used in food service to specify how to position a kitchen tool. For example, a recipe instruction may state, "hold the immersion blender at a 60° angle when puréeing the soup," or "cut the carrots on a bias by holding the knife at a 45° angle." **See Figure 3-20.**

Media Clips — Degrees and Angles

Using Angles

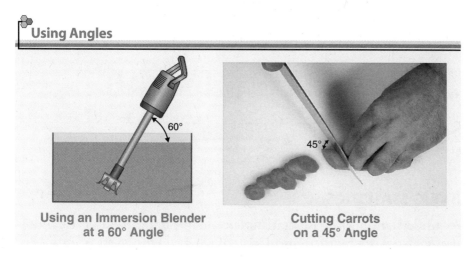

Using an Immersion Blender at a 60° Angle   Cutting Carrots on a 45° Angle

Figure 3-20. Angles can be used in recipe instructions to explain how to hold kitchen equipment when performing different tasks.

1. What is the definition of area?

_____

2. Stove top A measures 36 inches long by 26 inches wide. Stove top B measures 40 inches long by 24 inches wide. Which stove top has the larger area?

_____

_____

_____

3. How is the radius of a circle related to the diameter of a circle?

_____

4. What is the area of a round serving platter that measures 14 inches in diameter?

_____

_____

_____

5. What is the difference between volume and capacity?

_____

_____

6. A storage room measures 12 feet long by 6 feet wide by 8 feet high. What is the total capacity of the storage room in cubic feet?

_____

_____

_____

7. What is the capacity of a stockpot measuring 14 inches in diameter and 19 inches high? Round the answer to a whole number.

_____

_____

8. How many equal pieces has a circle been cut into if each piece has an angle of 90 degrees?

_____

_____

BASIC STATISTICS

An understanding of basic statistics is valuable in the foodservice industry. *Statistics* is a mathematical system for evaluating sets of numbers. The three most common statistical measurements are the average (or mean), median, and mode of a set of numbers.

An *average* is the sum of a set of numbers divided by how many numbers there are in the set. Another term for average is "mean." A *median* is the middle value within a set of numbers that are arranged in numerical order. Another way to think about the median is to picture a set of numbers where half of the set are numbers less than the median and the other half of the set are numbers greater than the median. The *mode* is the value that appears most frequently within a set of numbers.

The most common application of statistics in food service involves calculating averages such as the following:

- number of each menu item ordered per day
- number of customers per day
- number of hours worked per month per employee
- amount of a guest check
- amount of an operation's utility bills

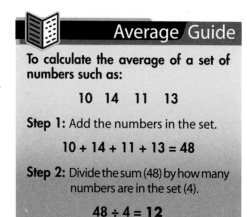

Average Guide

To calculate the average of a set of numbers such as:

10   14   11   13

Step 1: Add the numbers in the set.

10 + 14 + 11 + 13 = 48

Step 2: Divide the sum (48) by how many numbers are in the set (4).

48 ÷ 4 = 12

Count is another statistical measurement used in the professional kitchen. For example, shrimp are packaged by count, or number per pound, for a given size. For example, a bag of 16 - 20 count jumbo shrimp contains between 16 and 20 shrimp per pound. A 5-pound bag of 16 - 20 count colossal shrimp contains between 80 shrimp ($5 \times 16 = 80$) and 100 shrimp ($5 \times 20 = 100$).

If the shrimp found in nine 5-pound bags of 16 - 20 count shrimp were actually counted, the average, median, and mode of those numbers could be calculated. If the bags of shrimp each contain between 16 and 20 shrimp per pound, each 5-pound bag should contain between 80 and 100 shrimp ($16 \times 5 = 80$ and $20 \times 5 = 100$). Therefore, the average, median, and mode should all be between 80 and 100.

Creating a Data Table

When collecting a group of numbers, it is sometimes helpful to write them down in a table. A *data table* is a collection of information that is organized into a set of rows and columns. The top row of each column contains a heading that describes the data in the rows of that column. **See Figure 3-21.**

Data Tables

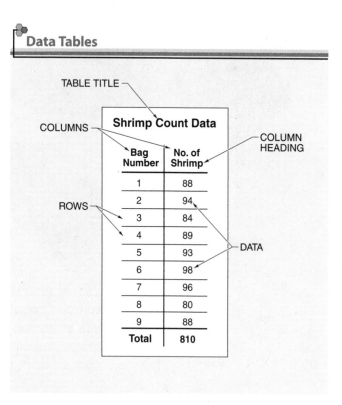

Figure 3-21. A data table is a good tool for organizing sets of numerical information.

According to the information shown in the data table, the total of shrimp counted is 810 and the total number of bags opened is 9. The total number of shrimp is required in order to calculate the average number of shrimp in each bag. In this case, there is an average of 90 shrimp per bag in a 5-pound bag.

810 shrimp ÷ 9 bags = 90 shrimp per bag

To find the median of the number of shrimp per bag, the data in the column for number of shrimp are listed in order from lowest to highest.

80   84   88   88   (89)   93   94   96   98

The median is 89 because it is the value that falls between the lower half of numbers (80, 84, 88, and 88) and the upper half of numbers (93, 94, 96, and 98). The mode is 88 shrimp since it is the most common number in the group.

By reviewing these statistics, it is confirmed that a 5-pound bag of 16 to 20 count shrimp from this seafood vendor will contain an average of 90 shrimp. Also, about half the bags will contain less than 89 shrimp (the median) and the other half will contain more than 89 shrimp. It will be most common to find 88 shrimp (the mode) in any one bag.

Harbor Seafood

Creating a Graph

When dealing with a large group of numbers, it may be easier to interpret data by looking at it in a graph rather than a table. A graph consists of an x-axis (horizontal axis), a y-axis (vertical axis), and data. In the shrimp example, the two sets of data are the individual bag numbers and the number of shrimp per bag. To represent data from the table in a graph, the bag numbers are placed along the x-axis and the counts of shrimp are placed along the y-axis.

To make the graph easy to read, the smallest number on each axis should be slightly less than the lowest data value in the table. Likewise, the highest number on each axis should be slightly more than the highest data value in the table. Since the smallest bag number is 1 and the highest bag number is 9, a range of 0 to 10 would be used for the x-axis. Since the smallest shrimp count is 80 and the highest is 98, a range of 76 to 100 would be used for the y-axis.

Data points are placed on the graph to match the information in the data table. For example, Bag 1 has a count of 88, so a data point is placed on the graph where the number 1 on the x-axis intersects with number 88 on the y-axis. Bag 2 has a count of 94, so a data point is placed where the number 2 on the x-axis intersects with number 94 on the y-axis, and so forth. Lines are drawn to connect the data points. The average (mean), median, and mode are calculated and represented by horizontal lines. **See Figure 3-22.**

Graphs

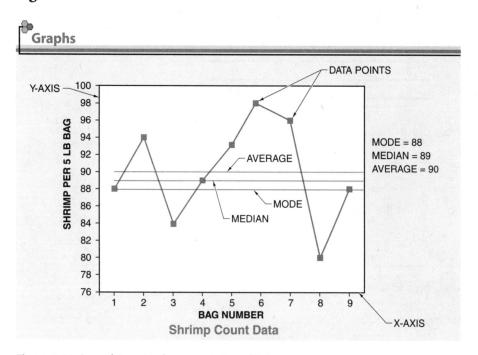

Figure 3-22. A graph is a visual representation of data.

Checkpoint 3-5

Master Math™ Applications

1. Explain how to calculate a statistical average.

_____

_____

2. Define median.

_____

_____

3. Define mode.

_____

_____

Use the data table shown to answer the following questions. The number of tortilla chips were counted in 11 different 1.5-ounce bags of tortilla chips.

Bag Number	Number of Chips	Bag Number	Number of Chips
1	12	7	18
2	14	8	17
3	16	9	19
4	22	10	20
5	16	11	18
6	16	**Total**	**188**

4. What is the average number of tortilla chips found in the 11 bags? Round to a whole number.

_____

_____

5. What is the median number of tortillas chips found in the 11 bags?

_____

_____

6. What is the mode of tortilla chips found in the 11 bags?

_____

Chapter 3 / Summary

Special math rules and formulas are required to perform calculations that consist of fractions and decimals. Fractions can take the form of proper fractions, improper fractions, and mixed numbers. Improper fractions and mixed numbers can be converted from one to the other as well as decimals and fractions. Solutions to decimal calculations are often rounded depending on the degree of accuracy required.

Measurements can also be used to calculate area and volume. Area is the size of a surface. Volume is the amount of space an object occupies or the capacity an object can hold. An angle is another form of measurement that is useful for describing how to portion an item or orient a piece of kitchen equipment. Measurements also include counts that can be used to generate statistical data for calculating averages, compiling a data table, or plotting a graph.

Detecto, A Division of Cardinal Scale Manufacturing Co.

Chapter **3**

Checkpoint Answers

Checkpoint 3-1

1. A measurement is a number with a corresponding unit that represents a size, capacity, amount, or quantity.

2. 14 c   (4 pt = 8 c, and 8 c + 6 c = 14 c)

3. 1 qt   (2 pt = 1 qt, and 2 qt – 1 qt = 1 qt)

4. 24 c   (4 c × 6 = 24 c)

Checkpoint 3-2

1. A fraction is a part of a whole.

2. The numerator represents the part and the denominator represents the whole.

3. A mixed number is a combination of a whole number and a fraction.

4. 8⅓   (25 ÷ 3 = 8 remainder 1)

5. ²¹⁄₈   [the numerator = (2 × 8) + 5 = 16 + 5 = 21, and the denominator stays the same]

6. ¾   (both the numerator and denominator have a factor of 3)

7. 25 c   (12½ = ²⁵⁄₂, ²⁵⁄₂ × ²⁄₁ = ⁵⁰⁄₂, and 50 ÷ 2 = 25)

8. 29 portions   (9⅔ = ²⁹⁄₃, and ²⁹⁄₃ ÷ ⅓ = ²⁹⁄₃ × ³⁄₁ = 29)

Checkpoint 3-3

1. A decimal is another way to depict a fraction that has a denominator that is a power of 10.

2. Rounding is the process of reducing the number of places in a decimal to achieve a certain degree of accuracy.

3. A decimal rounded to the one hundredths place.

4. Twelve and fifty-six hundredths.

5. 6.2

6. 34.658

7. 24.3 lb

8. 1.4 lb   (15.2 lb – 13.8 lb = 1.4 lb)

9. 45.36 c   (12.96 c per can × 3.5 cans = 45.36 c)

10. 25 portions   (62.5 ÷ 2.5 = 25)

11. ⅜   (both numerator and denominator have a factor of 125)

12. 0.67   (2 ÷ 3 = 0.6666)

continued . . .

Checkpoint Answers (continued)

Checkpoint 3-4

1. Area is the size of a two-dimensional (flat) object.

2. Stovetop B   (36 in. × 26 in. = 936 sq in. and 40 in. × 24 in. = 960 sq in.)

3. The radius of a circle is equal to one half the diameter.

4. 153.86 sq in.   [$r = 14 \div 2 = 7$, and $A = 3.14 \times (7 \text{ in})^2 = 3.14 \times 49 \text{ in}^2 = 153.86 \text{ in}^2$]

5. Volume is the size of a three-dimensional object. Capacity represents the volume of a substance that can be placed inside an empty three-dimensional object.

6. 576 cu ft   (12 ft × 6 ft × 8 ft = 576 cu ft)

7. 51 qt   [$3.14 \times (7 \text{ in.})^2 \times 19 \text{ in.} = 2923.34$ cu in., and $2923.34 \div 57.8 = 50.58$]

8. 4   ($360 \div 90 = 4$)

Checkpoint 3-5

1. An average is the sum of a group of numbers divided by how many numbers in the group.

2. A median is the middle value within a group of values that are arranged in numerical order.

3. Mode is the value that appears most frequently within a group of values.

4. 17   ($188 \div 11 = 17.09 = 17$ rounded to the nearest whole number)

5. 17   [half the numbers (12, 14, 16, 16, and 16) are less than 17 and half the numbers (18, 18, 19, 20, and 22) are greater than 17]

6. 16   (16 is the most common number, which appears 3 times in the set of numbers)

Converting Measurements and Scaling Recipes

R ecipes are often changed to produce more or less food to meet the demands of a particular kitchen. Foodservice employees need to be able to calculate new ingredient measurements to account for those changes in recipes. Often, measurements may need to be converted to different units of measure. Factors such as cooking times and temperatures also need to be taken into account when recipes are changed. By using solid math skills and paying close attention to detail, accurate changes can be made to recipes while maintaining the quality of the food prepared.

Chapter 4

Chapter Objectives

1. Convert measurements within the customary or metric measurement system.

2. Convert measurements between the customary and metric measurement systems.

3. Convert between volume and weight measurements.

4. Identify the most common elements of a standardized recipe.

5. Calculate scaling factors based on recipe yield.

6. Calculate scaling factors based on product availability.

7. Explain how other elements of a recipe are affected when the yield is changed.

Key Terms

- converting
- cancelling
- scaling
- yield
- portion size
- scaling factor

CONVERTING MEASUREMENTS

With a working knowledge of the standard units of measure used in the professional kitchen and an understanding of how measurements are calculated, foodservice employees can convert measurements. *Converting* is the process of changing a measurement with one unit of measure to an equivalent measurement with a different unit of measure. There are three different types of measurement conversions performed in the profession kitchen.

- **Converting volume or weight measurements within the customary or metric system.** Converting gallons to quarts, pounds to ounces, liters to milliliters, and grams to kilograms are all examples of this type of conversion.

- **Converting between customary and metric measurements.** Converting quarts to liters and pounds to grams are examples of this type of conversion. These conversion calculations require the use of conversion factors that are not whole numbers.

- **Converting between volume and weight measurements.** Converting cups to ounces, gallons to pounds, or teaspoons to grams are all examples of this type of conversion. These conversions are unique because volume-to-weight equivalents are approximations and differ depending on the ingredient being measured.

A common method for performing all of these measurement conversions involves expressing measurements and equivalents as fractions and cancelling the matching units. *Cancelling* is the process of crossing out and eliminating matching units in the numerators and denominators of fractions in a conversion calculation.

Carlisle FoodService Products

Converting Within the Customary or Metric System

To convert a measurement to a different unit within the same measurement system, the measurement is first written as a fraction. For example, to convert 8 quarts to gallons, the first step is to write the original measurement as the numerator in a fraction and 1 as the denominator.

$$\frac{8 \text{ qt}}{1}$$

The next step is to identify the conversion factor that the original measurement can be multiplied by to convert quarts to gallons. In this case, the common equivalent of 4 quarts = 1 gallon can be used. The equivalent can be written as a fraction in two ways, just like it is equally correct to say that "1 gallon is equivalent to 4 quarts" or "4 quarts are equivalent to 1 gallon." In either case the value of the equivalent fraction is 1.

$$\frac{1 \text{ gal.}}{4 \text{ qt}} \quad \text{or} \quad \frac{4 \text{ qt}}{1 \text{ gal.}}$$

The equivalent is written so that the unit in the denominator is the same as the unit in the original measurement. Then, the original measurement is multiplied by the equivalent.

$$\frac{8 \text{ qt}}{1} \times \frac{1 \text{ gal.}}{4 \text{ qt}}$$

When multiplied, the unit in the numerator of the first part of the calculation cancels the matching unit in the denominator in the next part of the calculation. Cancelling is shown by drawing a line through the matching units. In this example, the quarts cancel each other.

$$\frac{8 \cancel{\text{ qt}}}{1} \times \frac{1 \text{ gal.}}{4 \cancel{\text{ qt}}} = \frac{8}{1} \times \frac{1 \text{ gal.}}{4}$$

When multiplying fractions, the numerators are multiplied by each other and the denominators are multiplied by each other.

$$\frac{8 \times 1 \text{ gal.}}{1 \times 4} = \frac{8 \text{ gal.}}{4}$$

Then, the numerator is divided by the denominator to obtain the final answer.

8 gal. ÷ 4 = **2 gal.**

This example can also be shown as one continuous calculation.

$$\frac{8 \cancel{\text{ qt}}}{1} \times \frac{1 \text{ gal.}}{4 \cancel{\text{ qt}}} = \frac{8 \times 1 \text{ gal.}}{1 \times 4} = \frac{8 \text{ gal.}}{4} = 8 \text{ gal.} \div 4 = \textbf{2 gal.}$$

The cancelling method of conversion works for any measurement conversion within the same measurement system as long as the appropriate conversion factor is known. For example, 20 ounces of chocolate can be converted to pounds of chocolate. **See Figure 4-1.**

Converting Between Customary and Metric Measurements

Measurements can also be converted between customary and metric units of measure. However, the equivalents between customary units and metric units are not often whole numbers. For example, 1 liter is equivalent to 1.06 quarts and 1 kilogram is equivalent to 2.2 pounds. **See Figure 4-2.**

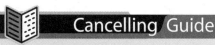

Cancelling Guide

Convert 4 pints to quarts.

Step 1: Write the measurement and the conversion factor as fractions in a calculation using multiplication.

$$\frac{4 \text{ pt}}{1} \times \frac{1 \text{ qt}}{2 \text{ pt}}$$

Step 2: Cancel the matching units in the numerators and denominators and then multiply the resulting fractions.

$$\frac{4 \cancel{\text{ pt}}}{1} \times \frac{1 \text{ qt}}{2 \cancel{\text{ pt}}} = \frac{4}{1} \times \frac{1 \text{ qt}}{2} = \frac{4 \text{ qt}}{2}$$

Step 3: Divide the numerator by the denominator to obtain the final answer.

4 qt ÷ 2 = 2 qt

Wisconsin Milk Marketing Board, Inc.

Forms and Tables

Example: Convert 20 ounces of chocolate to pounds of chocolate.

1. Write the original measurement as a fraction with the measurement as the numerator and 1 as the denominator.

ORIGINAL MEASUREMENT
$$\frac{20\ oz}{1}$$

2. Identify the equivalent between the original measurement and the desired measurement and write as a fraction with the original measurement unit in the denominator.

TO CONVERT OUNCES TO POUNDS
$$\frac{1\ lb}{16\ oz}$$

3. Set up a multiplication calculation with the measurement and the equivalent. Cancel the matching units in the numerators and denominators of the calculation.

$$\frac{20\ \cancel{oz}}{1} \times \frac{1\ lb}{16\ \cancel{oz}}$$

CANCEL MATCHING UNITS

4. Multiply the resulting fractions.

$$\frac{20}{1} \times \frac{1\ lb}{16} = \frac{20 \times 1\ lb}{1 \times 16} = \frac{20\ lb}{16}$$

5. Divide the numerator by the denominator to obtain the final answer.

$$20\ lb \div 16 = 1.25\ lb$$

Answer: 1.25 lb

CONVERSION SHOWN AS ONE CONTINUOUS CALCULATION

$$\frac{20\ \cancel{oz}}{1} \times \frac{1\ lb}{16\ \cancel{oz}} = \frac{20}{1} \times \frac{1\ lb}{16} = \frac{20\ lb}{16} = 20\ lb \div 16 = 1.25\ lb$$

Figure 4-1. Converting measurements involves expressing measurements and equivalents as fractions and cancelling the matching units of measure.

Customary and Metric Unit Equivalents

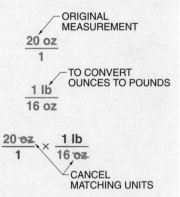

Volume
1 gallon (gal.) = 3.79 liters (L)
1 quart (qt) = 0.95 liters (L)
1 cup = 236.6 milliliters (mL)
1 fluid ounce (fl oz) = 29.6 milliliters (mL)
1 teaspoon (tsp) = 5 milliliters (mL)
1 liter (L) = 1.06 quart (qt)
1 liter (L) = 33.8 fluid ounces (fl oz)

1 LITER IS A LITTLE LARGER THAN A QUART

1 liter — 1 quart

500 milliliters —

250 milliliters — 1 cup

Weight
1 pound (lb) = 0.454 killograms (kg)
1 pound (lb) = 454 grams (g)
1 ounce (oz) = 28.4 grams (g)
1 kilogram (kg) = 2.2 pounds (lb)

1 pound (454 grams)

1 kilogram (2.2 pounds)

1 KILOGRAM IS A LITTLE LARGER THAN TWO POUNDS

Figure 4-2. Equivalents between customary and metric system units of measure for volume and weight are used to perform calculations in the professional kitchen.

Although the customary to metric equivalents may not be as easy to remember as the equivalents within the same system of measurement, the process for converting these measurements is exactly the same as other conversions. For example, to convert 800 grams to pounds (metric units to customary units) the first step is to write 800 grams as a fraction with the measurement in the numerator and 1 in the denominator.

$$\frac{800\ g}{1}$$

The second step is to identify the equivalent between grams and pounds. The equivalent 1 pound = 454 grams is found in Figure 4-2. This equivalent is written as a fraction with grams in the denominator so that the grams will cancel each other.

$$\frac{800\ \cancel{g}}{1}\times\frac{1\ lb}{454\ \cancel{g}}=\frac{800}{1}\times\frac{1\ lb}{454}$$

Next, the numerators of each fraction are multiplied by each other and the denominators of each fraction are multiplied by each other.

$$\frac{800\times 1\ lb}{1\times 454}=\frac{800\ lb}{454}$$

Then, the numerator is divided by the denominator to obtain the final answer.

$$800\ lb \div 454 = 1.76\ lb$$

This conversion can also be written as one continuous calculation.

$$\frac{800\ \cancel{g}}{1}\times\frac{1\ lb}{454\ \cancel{g}}=\frac{800\times 1\ lb}{1\times 454}=\frac{800\ lb}{454}=800\ lb \div 454 = \mathbf{1.76\ lb}$$

Note: When converting between customary and metric measurements, it is common to calculate answers as decimals. Decimal answers should be rounded based on the degree of accuracy required for that ingredient in the recipe.

In some instances, two measurement equivalents will need to be used in one conversion calculation. For example, if a restaurant has 2 liters of soy sauce in storage and a recipe requires 3 pints of soy sauce, is there enough soy sauce to make the recipe? To answer this question, two equivalents are required if a single equivalent between liters and pints is not provided. The first equivalent is used to convert liters to quarts and then the second equivalent is used to convert quarts to pints. **See Figure 4-3.**

Example: If a restaurant has 2 liters of soy sauce in storage and a recipe requires 3 pints of soy sauce, is there enough soy sauce to make the recipe?

1. Write the original measurement as a fraction with the original measurement as the numerator and 1 as the denominator.

ORIGINAL MEASUREMENT

$$\frac{2\ L}{1}$$

2. Identify the equivalents between the original measurement and the desired measurement and write as fractions with the original measurement unit in the denominator.

TO CONVERT LITERS TO QUARTS

$$\frac{1\ qt}{0.95\ L} \quad \text{and} \quad \frac{2\ pt}{1\ qt}$$

TO CONVERT QUARTS TO PINTS

3. Set up a multiplication calculation with the measurement and the equivalents. Cancel the matching units in the numerators and denominators of the calculation.

CANCEL MATCHING UNITS

$$\frac{2\ L}{1} \times \frac{1\ qt}{0.95\ L} \times \frac{2\ pt}{1\ qt}$$

CANCEL MATCHING UNITS

4. Multiply the resulting fractions.

$$\frac{2}{1} \times \frac{1}{0.95} \times \frac{2\ pt}{1} = \frac{4\ pt}{0.95}$$

5. Divide the numerator by the denominator to obtain the final answer (rounded to the tenths place).

$$4\ pt \div 0.95 = 4.2\ pt$$

Answer: Yes, there are 4.2 pints.

CONVERSION SHOWN AS ONE CONTINUOUS CALCULATION

$$\frac{2\ L}{1} \times \frac{1\ qt}{0.95\ L} \times \frac{2\ pt}{1\ qt} = \frac{2}{1} \times \frac{1}{0.95} \times \frac{2\ pt}{1} = \frac{4\ pt}{0.95} = 4\ pt \div 0.95 = 4.2\ pt$$

Figure 4-3. Volume measurements in metric units can be converted to volume measurements in customary units.

Converting Between Volume and Weight Measurements

Converting between volume and weight measurements is often necessary because food products cannot always be purchased in the same units that are called for in a recipe. **See Figure 4-4.** Equivalents between volume and weight units are approximations based on how much a given volume of an ingredient will weigh on a scale. Two ingredients of the same volume may not weigh the same due to differences in density. For example, 8 fluid ounces of water will weigh 8 ounces, but 8 fluid ounces of flour will only weigh about 4.5 ounces, while 8 fluid ounces of honey will weigh about 12 ounces. Since it would not be reasonable to memorize the volume-to-weight equivalents for all the ingredients used in the professional kitchen, equivalent tables are referenced as needed. **See Appendix.**

It is important to remember that volume-to-weight equivalents are approximate and care must be taken to reference the most appropriate entry in an equivalents table. For example, there may be more than one volume-to-weight equivalent listed for grapes depending on whether the grapes are sliced or if the grapes are whole. A cup of sliced grapes will weigh more than a cup of whole grapes because the sliced grapes will fill the cup more efficiently. **See Figure 4-5.**

Volume vs. Weight Units

Blueberry Pie

Yield: 1 (9-in. pie)

Ingredients

24 oz	Blueberries, fresh
6 oz	sugar, granulated
6 fl oz	water
2 tsp	lemon juice

Recipe Calls for Ounces   **Product Sold in Pints**

Figure 4-4. Food products are not always sold in the same units as the units used in a recipe.

Volume-to-Weight Equivalents

Volume-to-Weight Table		
Ingredient	*Volume*	*Weight*
Grapes		
sliced	1 c	5¾ oz
whole	1 c	3¾ oz

1 Cup Grapes, Sliced   **1 Cup Grapes, Whole**

Figure 4-5. When referencing a table for volume-to-weight equivalents, there may be more than one entry for an ingredient based on its form.

Volume-to-weight equivalents can be used to convert volume measurements to weight measurements. For example, how many ounces of frozen peas are there in 10 cups of frozen peas? The first step is to write the original measurement of 10 cups as a fraction.

Media Clips — Volume-to-Weight Equivalents

$$\frac{10\,c}{1}$$

Based on the equivalents table, the volume-to-weight equivalent for frozen peas is 1 cup = 3.5 ounces. This equivalent is written with cups in the denominator so the cups will cancel each other.

Forms and Tables

$$\frac{10\,c}{1} \times \frac{3.5\,oz}{1\,c} = \frac{10}{1} \times \frac{3.5\,oz}{1}$$

The numerators of each fraction and denominators of each fraction are then multiplied by each other. Then, the numerator is divided by the denominator to obtain the final answer.

$$\frac{10 \times 3.5 \text{ oz}}{1 \times 1} = \frac{35 \text{ oz}}{1} = 35 \text{ oz}$$

This conversion can also be written as one continuous calculation.

$$\frac{10 \cancel{\text{ c}}}{1} \times \frac{3.5 \text{ oz}}{1 \cancel{\text{ c}}} = \frac{10 \times 3.5 \text{ oz}}{1 \times 1} = \frac{35 \text{ oz}}{1} = 35 \text{ oz} \div 1 = \mathbf{35 \text{ oz}}$$

Volume-to-weight equivalents can also be used to convert weight measurements to volume measurements. Consider the example of a recipe that requires 24 ounces of fresh blueberries. If the blueberries are only available for purchase by the pint, how many pints of blueberries should be ordered? To answer this question, 24 ounces of blueberries is converted to pints. **See Figure 4-6.**

Converting Between Volume and Weight Measurements

Example: A recipe requires 24 ounces of fresh blueberries and fresh blueberries are purchased in 1-pint containers. How many pints of blueberries should be ordered?

Tip: 1 cup of fresh blueberries weighs 7 ounces.

1. Write the original measurements as a fraction with the measurement as the numerator and 1 as the denominator.

$$\frac{24 \text{ oz}}{1} \quad \text{ORIGINAL MEASUREMENT}$$

2. Identify the equivalents between the original measurement and the desired measurement and write as fraction with the original measurement unit in the denominator.

$$\frac{1 \text{ c}}{7 \text{ oz}} \text{ (TO CONVERT OUNCES TO CUPS)} \quad \text{and} \quad \frac{1 \text{ pt}}{2 \text{ c}} \text{ (TO CONVERT CUPS TO PINTS)}$$

3. Set up a multiplication calculation with the original measurement and the equivalents. Cancel the matching units in the numerators and denominators of the calculation.

$$\frac{24 \cancel{\text{ oz}}}{1} \times \frac{1 \cancel{\text{ c}}}{7 \cancel{\text{ oz}}} \times \frac{1 \text{ pt}}{2 \cancel{\text{ c}}} \quad \text{CANCEL MATCHING UNITS}$$

4. Multiply the resulting fractions.

$$\frac{24}{1} \times \frac{1}{7} \times \frac{1 \text{ pt}}{2} = \frac{24 \text{ pt}}{14}$$

5. Divide the numerator by the denominator to obtain the final answer.

$$24 \text{ pt} \div 14 = \mathbf{1.7 \text{ pt}}$$

Answer: Since 1.7 pints are required, 2 pints should be ordered.

$$\frac{24 \cancel{\text{ oz}}}{1} \times \frac{1 \cancel{\text{ c}}}{7 \cancel{\text{ oz}}} \times \frac{1 \text{ pt}}{2 \cancel{\text{ c}}} = \frac{24}{1} \times \frac{1}{7} \times \frac{1 \text{ pt}}{2} = \frac{24 \text{ pt}}{14} = 24 \text{ pt} \div 14 = \mathbf{1.7 \text{ pt}}$$

CONVERSION SHOWN AS ONE CONTINUOUS CALCULATION

Figure 4-6. Weight measurements can be converted to volume measurements using approximate volume-to-weight equivalents.

1. Define cancelling.

_____

_____

2. Convert 6 cups to quarts.

_____

_____

3. Convert 4.25 kilograms to grams.

_____

_____

4. Convert 10 cups to liters. Round answer to the tenths place. (*Tip: 1 qt = 0.95 L*)

_____

_____

5. Convert 40 ounces to grams. (*Tip: 28.4 g = 1 oz*)

_____

_____

6. Why would it be important to distinguish between chopped pecans and whole pecans when looking up a volume-to-weight equivalent for pecans in a table?

_____

_____

7. If 1 cup of fresh raspberries weighs 2.75 ounces, how many cups are in 2 pounds of fresh raspberries? Round answer to the tenths place.

_____

_____

SCALING RECIPES

Converting measurements is often done when scaling recipes. *Scaling* is the process of calculating new amounts for each ingredient in a recipe when the total amount of food the recipe makes is changed. For example, a recipe that serves 4 people may be scaled for use in a restaurant that plans to make 50 servings of the recipe. Similarly, a recipe used by a banquet hall that normally serves 100 people may need to be scaled to make only 12 servings for a small party.

A recipe may also need to be scaled based on the availability of one of the ingredients. For example, a recipe may require 20 pounds of ground beef but only 15 pounds of ground beef are available. A foodservice employee needs to know how to adjust all of the other ingredient amounts in a recipe to account for the available amount of a key ingredient.

Foodservice employees not only need to know how to calculate new ingredient amounts when recipes are scaled, but also need to understand how scaling can affect other elements of a recipe. For example, depending on how significantly the recipe is scaled, the units of measure used for some ingredients may need to be converted to make measuring the ingredients more practical. In addition, cooking times or temperatures may need to be adjusted, or cooking equipment of a different size may be required. A well-written recipe will contain all of the information that needs to be considered when a recipe is scaled.

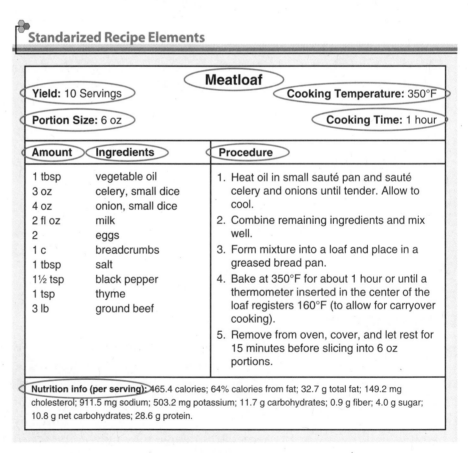

Forms and Tables

Standardized Recipe Elements

The exact look and format of a standardized recipe will vary from one foodservice operation to another. However, most standardized recipes usually contain the following common elements. **See Figure 4-7.**

Standarized Recipe Elements

Meatloaf

Yield: 10 Servings **Cooking Temperature:** 350°F

Portion Size: 6 oz **Cooking Time:** 1 hour

Amount	Ingredients	Procedure
1 tbsp	vegetable oil	1. Heat oil in small sauté pan and sauté celery and onions until tender. Allow to cool.
3 oz	celery, small dice	
4 oz	onion, small dice	
2 fl oz	milk	2. Combine remaining ingredients and mix well.
2	eggs	
1 c	breadcrumbs	3. Form mixture into a loaf and place in a greased bread pan.
1 tbsp	salt	
1½ tsp	black pepper	4. Bake at 350°F for about 1 hour or until a thermometer inserted in the center of the loaf registers 160°F (to allow for carryover cooking).
1 tsp	thyme	
3 lb	ground beef	
		5. Remove from oven, cover, and let rest for 15 minutes before slicing into 6 oz portions.

Nutrition info (per serving): 465.4 calories; 64% calories from fat; 32.7 g total fat; 149.2 mg cholesterol; 911.5 mg sodium; 503.2 mg potassium; 11.7 g carbohydrates; 0.9 g fiber; 4.0 g sugar; 10.8 g net carbohydrates; 28.6 g protein.

Figure 4-7. Most standardized recipes include the same common elements.

- **Recipe Name.** The name of a recipe should be descriptive of the dish being prepared and should reflect the name used on the menu. Recipes should be named to avoid confusion in the kitchen. For example, if a restaurant makes two kinds of chili sauce (a spicy version and a mild version) both recipes should not be named "Chili Sauce." Instead, more descriptive names, such as Spicy Red Chili Sauce and Mild Green Chili Sauce, should be used.

- **Yield.** *Yield* is the total quantity of a food or beverage item that is made from a standardized recipe. Yields can be expressed as a count, a total volume or weight, or a number of portions. For example, 24 cookies, 3 gallons, 50 pounds, and 36 8-ounce portions are all valid recipe yields.

- **Portion Size.** *Portion size* is the amount of a food or beverage item that is served to an individual person. Portion size is related to yield. For example, a soup recipe that yields 1 gallon can also be said to yield 16 portions of 8 fluid ounces each because 16×8 fluid ounces = 128 fluid ounces = 1 gallon. **See Figure 4-8.**

- **Ingredients and Procedures.** The amount of each ingredient used in the recipe is listed next to the name of the ingredient. If an ingredient is to be prepared in a certain way prior to being measured, such as minced or chopped, that information is also provided. The ingredients are usually listed in the order that they are incorporated into the recipe to help ensure that no ingredient is left out. Procedures are listed in sequential order.

- **Cooking Temperature.** The cooking temperature may be an exact temperature at which to set an oven or a more general indication of temperature such as "low heat" or "high heat." The temperature at which food is cooked greatly affects the outcome of the final product. For example, food that is cooked at too high of a temperature may burn on the outside before it is properly cooked in the center.

- **Cooking Time.** Cooking times provided in standardized recipes are often treated as guidelines. However, professional cooks with experience rely more on the appearance or feel of an item than on a cooking time. They also often rely on exact measurements, such as an internal temperature checked with a thermometer, to determine when a food item is done.

- **Nutrition Information.** While nutrition information is not required to prepare a recipe, it is important information to have for menu planning and for customer inquiries. Due to health or dietary concerns, customers may ask how much fat, carbohydrates, or sodium content there is in a menu item.

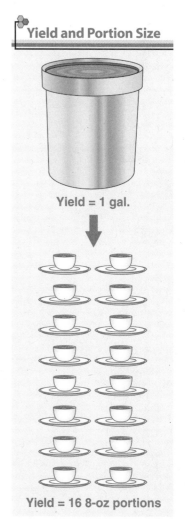

Yield and Portion Size

Yield = 1 gal.

Yield = 16 8-oz portions

Figure 4-8. A total recipe yield can be expressed as a number of portions and portion size.

Scaling Based on Yield

The most straightforward examples of scaling a recipe are doubling the yield where all the ingredient amounts are simply multiplied by 2, or halving the yield where all the ingredient amounts are simply divided by 2.

Scaling becomes more challenging when the yield is changed by a factor that is not as simple as doubling or halving the recipe. For example, a recipe that normally yields 3 gallons may need to be scaled to yield 10 gallons. Likewise, a recipe that makes 12 6-ounce portions may need to be scaled to make 30 9-ounce portions.

Regardless of the reason for scaling a recipe, the scaling process starts out by calculating a scaling factor. A *scaling factor* is the number that each ingredient amount in a recipe is multiplied by when the recipe yield is changed. The formula for calculating a scaling factor based on yield is as follows:

$$SF = DY \div OY$$

where
SF = scaling factor
DY = desired yield
OY = original yield

$$Scaling\ Factor = \frac{Desired\ Yield}{Original\ Yield}$$

Scaling Based on Count. If a cookie recipe that makes 24 cookies is scaled to make 84 cookies, the scaling factor is calculated by dividing the desired yield (84 cookies) by the original yield (24 cookies). **See Figure 4-9.**

$$SF = DY \div OY$$

SF = 84 cookies ÷ 24 cookies

SF = **3.5**

Scaling Based on Volume. If a soup recipe that makes 8 gallons of soup is scaled to make 3 gallons of soup, the scaling factor is calculated by dividing the desired yield (3 gallons) by the original yield (8 gallons).

$$SF = DY \div OY$$

SF = 3 gal. ÷ 8 gal.

SF = **0.375**

Scaling Based on Weight. If a potato salad recipe that makes 4 pounds of salad is scaled to make 80 pounds of salad, the scaling factor is calculated by dividing the desired yield (80 pounds) by the original yield (4 pounds).

$$SF = DY \div OY$$

SF = 80 lb ÷ 4 lb

SF = **20**

Scaling Based on Portion Size. If a fish recipe that makes 12 8-ounce portions is scaled to make 30 10-ounce portions, the yields must first be converted to a total number of ounces.

$OY = 12 \times 8 \text{ oz} = 96 \text{ oz}$

$DY = 30 \times 10 \text{ oz} = 300 \text{ oz}$

Then the scaling factor formula can be applied.

SF = DY ÷ OY

$SF = 300 \text{ oz} \div 96 \text{ oz}$

$SF = \mathbf{3.125}$

Media Clips — Calculating Scaling Factors

Calculating Scaling Factors

$$\text{Scaling Factor} = \frac{\text{Desired Yield}}{\text{Original Yield}}$$

Type of Yield	Original Yield	Desired Yield	Scaling Factor
Count Yield	24 cookies	84 cookies	84 ÷ 24 = **3.5**
Total Volume Yield	8 gallons	3 gallons	3 ÷ 8 = **0.375**
Total Weight Yield	4 pounds	80 pounds	80 ÷ 4 = **20**
Portion Yield	12 8-oz portions (12 × 8 oz = 96 oz)	30 10-oz portions (30 × 10 oz = 300 oz)	300 ÷ 96 = **3.125**

Figure 4-9. When a recipe is scaled based on yield, the scaling factor is calculated by dividing the desired yield by the original yield.

Scaling Based on Product Availability

In some cases, a recipe is scaled because the amount of one ingredient needs to be changed based on availability. In situations like this, the scaling factor is calculated by dividing the available amount of the ingredient by the original amount of the ingredient. The formula for calculating a scaling factor based on product availability is as follows:

SF = AA ÷ OA

where
SF = scaling factor
AA = available amount
OA = original amount

$$\text{Scaling Factor} = \frac{\text{Available Amount}}{\text{Original Amount}}$$

For example, if a beef stew recipe that calls for 15 pounds of beef stew meat needs to be made with only 12 pounds of beef stew meat, the scaling factor is calculated by dividing the available ingredient amount by the original ingredient amount.

SF = AA ÷ OA

SF = 12 lb beef stew meat ÷ 15 lb beef stew meat

SF = **0.8**

On the other hand, if there are 18 pounds of beef stew meat available and all of the meat is to be used, the scaling factor would be 1.2.

SF = AA ÷ OA

SF = 18 lb beef stew ÷ 15 lb beef stew

SF = **1.2**

The Beef Checkoff

Multiplying Scaling Factors

Regardless of the method used for calculating the scaling factor, the process of scaling the recipe is the same. Once a scaling factor is known, every ingredient amount in the original recipe is multiplied by the scaling factor. **See Figure 4-10.** Using these new ingredient amounts will produce the desired yield of the scaled recipe. For example, a meatloaf recipe that normally yields 10 servings is scaled to make 80 servings. The scaling factor (based on count) is calculated first.

SF = DY ÷ OY

SF = 80 servings ÷ 10 servings

SF = **8.0**

Then, the new amount of each ingredient is calculated using the following formula:

NA = OA × SF   *New Amount = Original Amount × Scaling Factor*

where
NA = new amount
OA = original amount
SF = scaling factor

For example, the original meatloaf recipe requires 3 pounds of ground beef. The new amount of ground beef required is calculated as follows:

NA = OA × SF

$NA = 3 \text{ lb} \times 8.0$

$NA = \textbf{24 lb}$

The new amounts for the remaining ingredients are then calculated using the same formula. It may be necessary to convert some of the new ingredient amounts to a different unit of measure to make the measuring of the ingredients more efficient.

In general, it is most efficient to measure ingredients using the largest appropriate unit of measure. For example, the original meatloaf recipe in Figure 4-10 calls for 1 tablespoon of vegetable oil. The new amount based on a scaling factor of 8 is calculate as follows:

The Spice House.com

NA = OA × SF

$NA = 1 \text{ tbsp} \times 8.0$

$NA = \textbf{8 tbsp}$

If the scaling factor were smaller, such as 2, the new ingredient amount would only be 2 tablespoons. Measuring 2 tablespoons of vegetable oil by measuring two individual tablespoons is easy. However, measuring 8 tablespoons one tablespoon at a time is not practical. Instead, 8 tablespoons are converted to a larger unit of measure such as cups. First, tablespoons are converted to ounces.

$$\frac{8 \text{ tbsp}}{1} \times \frac{1 \text{ fl oz}}{2 \text{ tbsp}} = \frac{8 \times 1 \text{ fl oz}}{1 \times 2} = \frac{8 \text{ fl oz}}{2} = 8 \text{ fl oz} \div 2 = 4 \text{ fl oz}$$

Then, fluid ounces are converted to cups.

$$\frac{4 \text{ fl oz}}{1} \times \frac{1 \text{ c}}{8 \text{ fl oz}} = \frac{4 \times 1 \text{ c}}{1 \times 8} = \frac{4 \text{ c}}{8} = 4 \text{ c} \div 8 = \textbf{0.5 c}$$

Measuring 0.5 cup (or ½ cup) is more efficient than measuring eight individual tablespoons.

Meatloaf Recipe Scaled from 10 Servings to 80 Servings Scaling Factor = 80 ÷ 10 = 8			
Yield = 10 Servings		**Yield = 80 Servings**	
Amount	*Ingredients*	*Amount*	*Ingredients*
❶ 1 tbsp	vegetable oil	½ c	vegetable oil
❷ 3 oz	celery, small dice	1 lb 8 oz	celery, small dice
❸ 4 oz	onion, small dice	2 lb	onion, small dice
❹ 2 fl oz	milk	1 pt	milk
❺ 2	eggs, beaten	16	eggs, beaten
❻ 1 c	breadcrumbs	2 qt	breadcrumbs
❼ 1 tbsp	salt	½ c	salt
❽ 1½ tsp	black pepper	4 tbsp	black pepper
❾ 1½ tsp	thyme	4 tbsp	thyme
❿ 3 lb	ground beef	24 lb	ground beef

❶ 1 tbsp × 8 = 8 tbsp: $\dfrac{8\text{ tbsp}}{1} \times \dfrac{1\text{ fl oz}}{2\text{ tbsp}} \times \dfrac{1\text{ c}}{8\text{ fl oz}} = \dfrac{8\text{ c}}{16} = $ **½ c**

❷ 3 oz × 8 = 24 oz: $\dfrac{24\text{ oz}}{1} \times \dfrac{1\text{ lb}}{16\text{ oz}} = \dfrac{24\text{ lb}}{16} = 1.5\text{ lb} = 1\tfrac{1}{2}\text{ lb} = $ **1 lb 8 oz**

❸ 4 oz × 8 = 32 oz: $\dfrac{32\text{ oz}}{1} \times \dfrac{1\text{ lb}}{16\text{ oz}} = \dfrac{32\text{ lb}}{16} = $ **2 lb**

❹ 2 fl oz × 8 = 16 fl oz: $\dfrac{16\text{ fl oz}}{1} \times \dfrac{1\text{ c}}{8\text{ fl oz}} \times \dfrac{1\text{ pt}}{2\text{ c}} = \dfrac{16\text{ pt}}{16} = $ **1 pt**

❺ 2 eggs × 8 = **16 eggs**

❻ 1 c × 8 = 8 c: $\dfrac{8\text{ c}}{1} \times \dfrac{1\text{ pt}}{2\text{ c}} \times \dfrac{1\text{ qt}}{2\text{ pt}} = \dfrac{8\text{ qt}}{4} = $ **2 qt**

❼ 1 tbsp × 8 = 8 tbsp: $\dfrac{8\text{ tbsp}}{1} \times \dfrac{1\text{ fl oz}}{2\text{ tbsp}} \times \dfrac{1\text{ c}}{8\text{ fl oz}} = \dfrac{8\text{ c}}{16} = $ **½ c**

❽ 1½ tbsp × 8 = 12 tbsp: $\dfrac{12\text{ tsp}}{1} \times \dfrac{1\text{ tbsp}}{3\text{ tsp}} = \dfrac{12\text{ tbsp}}{3} = $ **4 tbsp**

❾ 1½ tbsp × 8 = 12 tbsp: $\dfrac{12\text{ tsp}}{1} \times \dfrac{1\text{ tbsp}}{3\text{ tsp}} = \dfrac{12\text{ tbsp}}{3} = $ **4 tbsp**

❿ 3 lb × 8 = **24 lb**

Figure 4-10. When the quantities of a recipe ingredient are multiplied by a scaling factor, it may be necessary to adjust the new measurements to a more appropriate unit of measure.

Additional Scaling Considerations

When recipes are scaled, additional considerations must be taken into account before actually preparing the recipe.

- **Adjusting Measurements.** If a new measurement is calculated that is not easily measured, such as 3.7 cups, the measurement will need to be adjusted to make measuring more practical. However, to avoid affecting the final result, the amount should not be adjusted drastically. For example, 3.7 cups should be adjusted to 3¾ cups (3.75 cups).

- **Sizing Cooking Equipment.** Cooking equipment must properly accommodate the adjusted amount of food. It is important that cooking equipment is large enough to handle an increased yield. Care must also be taken when decreasing a yield because cooking too little of an ingredient in too large of a pot or pan results in the rapid evaporation of liquids that leads to scorching or burning.

- **Adjusting Cooking Time.** A professional cook must determine whether the cooking time for a recipe needs to be increased or decreased. If a larger roast is used for a recipe, a longer cooking time will normally be required. However, 10 steaks cooked on a grill will cook in the same amount of time as 2 steaks on a grill. If a cake recipe that normally makes four 3-inch thick round cake layers is scaled up and used to make one large 1-inch thick sheet cake, the thinner sheet of cake may cook faster than the thicker cake layers.

The Beef Checkoff

- **Adjusting Cooking Temperature.** A professional cook must determine whether the cooking temperature needs to be adjusted. Normally, cooking time is adjusted but in certain circumstances, such as when a recipe yield is significantly decreased, it may be necessary to reduce the cooking temperature to keep food from drying out.

- **Adjusting Mixing Time.** Instructions related to mixing times may need to be adjusted, especially when a recipe yield is significantly increased, to ensure that the appropriate amount of time is given to adequately mix the ingredients.

Addressing these considerations will help to ensure that the final product is of the same quality as the original recipe.

1. Define yield.

   _____

   _____

2. Define portion size.

   _____

   _____

3. What is the yield of a recipe in quarts that makes 8 4-fluid ounce portions?

   _____

   _____

4. Define scaling factor.

   _____

   _____

5. If a standardized recipe that yields 2 quarts of soup is scaled to yield 2 gallons, what is the scaling factor?

   _____

   _____

6. If a standardized recipe that yields 20 6-ounce portions of stew is scaled to yield 10 8-ounce portions, what is the scaling factor?

   _____

   _____

7. If a standardized recipe for chicken pot pie calls for 10 pounds of cooked chicken meat and is scaled to use 14 pounds of available cooked chicken meat, what is the scaling factor?

   _____

   _____

8. If a standardized recipe calls for 3 ounces of cheddar cheese and the recipe is scaled by a scaling factor of 16, how many pounds of cheese are required?

   _____

   _____

9. Explain why a piece of equipment called for in a recipe may need to be changed due to its size when the recipe is scaled.

   _____

   _____

10. Give an example of when the cooking time for a recipe might need to be changed because the recipe has been scaled.

_____

_____

Chapter 4 Summary

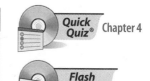

Quick Quiz® Chapter 4

Flash Cards

Measurement conversions are made using the same process whether converting within volume or weight units, converting between customary and metric units, or converting between volume and weight units. This process involves expressing measurements and equivalents as fractions and the cancelling of matching units.

Standardized recipes include many common elements, including the recipe yield. A recipe yield can be expressed as a count, a total volume or weight, or a number of portions. Often, the yield of a recipe needs to be increased or decreased. This process is referred to as scaling. When a recipe is scaled, new ingredient amounts must be calculated based on a scaling factor. Other elements of the recipe, such as cooking times or temperatures, must also be evaluated. The process of converting measurements and scaling recipes requires the accurate application of math skills in order to ensure that the quality of the final product is maintained.

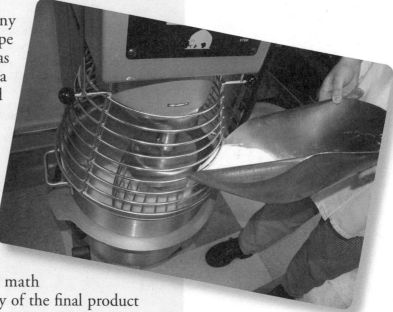

Checkpoint Answers

1. Cancelling is the process of crossing out and eliminating matching units or factors in the numerators and denominators of fractions in a conversion calculation.

2. 1.5 qt $\left(\dfrac{6\,\cancel{c}}{1}\times\dfrac{1\,qt}{4\,\cancel{c}}=\dfrac{6\times1\,qt}{1\times4}=\dfrac{6\,qt}{4}=6\,qt\div4=1.5\,qt\right)$

3. 4250 g $\left(\dfrac{4.25\,\cancel{kg}}{1}\times\dfrac{1000\,g}{1\,\cancel{kg}}=\dfrac{4.25\times1000\,g}{1\times1}=\dfrac{4250\,g}{1}=4.25\,g\right)$

4. 2.4 qt $\left(\dfrac{10\,\cancel{c}}{1}\times\dfrac{1\,\cancel{qt}}{4\,\cancel{c}}\times\dfrac{0.95\,L}{1\,\cancel{qt}}=\dfrac{10\times1\times0.95\,L}{1\times4\times1}=\dfrac{9.5\,qt}{4}=9.5\,qt\div4=2.4\,qt\right)$

5. 1136 g $\left(\dfrac{40\,\cancel{oz}}{1}\times\dfrac{28.4\,g}{1\,\cancel{oz}}=\dfrac{40\times28.4\,g}{1\times1}=\dfrac{1136\,g}{1}=1136\,g\right)$

6. The volume-to-weight equivalent for chopped pecans will be higher than the equivalent for whole pecans because the chopped pecans will fill a volume measurement tool more efficiently.

7. 11.6 c $\left(\dfrac{2\,\cancel{lb}}{1}\times\dfrac{16\,\cancel{oz}}{1\,\cancel{lb}}\times\dfrac{1\,c}{2.75\,\cancel{oz}}=\dfrac{2\times16\times1\,c}{1\times1\times2.75}=\dfrac{32\,c}{2.75}=32\,c\div2.75=11.6\,c\right)$

1. Yield is a term used to describe the total quantity of a food or beverage product that is made from following a standardized recipe.

2. Portion size is the amount of a food or beverage item that is served to an individual person.

3. 1 qt $\left(8\times4\,fl\,oz=32\,fl\,oz\text{ and }\dfrac{32\,\cancel{fl\,oz}}{1}\times\dfrac{1\,qt}{32\,\cancel{fl\,oz}}=\dfrac{32\,qt}{32}=1\,qt\right)$

4. A scaling factor is the number that all ingredient amounts are multiplied by when the yield of a recipe is changed.

5. 4 $\left(\dfrac{2\,\cancel{gal.}}{1}\times\dfrac{4\,qt}{1\,\cancel{gal.}}=8\,qt\text{ and }8\,qt\div2\,qt=4\right)$

6. 0.67 (20 × 6 oz = 120 oz; 10 × 8 oz = 80 oz; and 80 oz ÷ 120 oz = 0.67)

7. 1.4 (14 ÷ 10 = 1.4)

8. 3 lb $\left(3\,oz\times16=48\,oz\text{ and }\dfrac{48\,\cancel{oz}}{1}\times\dfrac{1\,lb}{16\,\cancel{oz}}=3\,lb\right)$

9. Equipment must be the correct size to hold a larger amount of food or to avoid cooking too small an amount of food in too large of a pot or pan.

10. Cooking time may need to be adjusted when the size or thickness of the food item being prepared changes.

Calculating Percentages and Ratios

Foodservice workers use percentages to determine how much of a given food product is usable. They use those percentages when processing food orders and planning for production. Many food products must be cleaned and/or trimmed before they can be used in a recipe, cooked, or served. Meats are trimmed of excess fat, fish are deboned and filleted, and many fruits and vegetables are peeled and seeded. In each case, only a portion of the food product is usable and what remains becomes waste. Percentages, as well as ratios, make recipes easier to use and help guarantee consistent results.

Chapter Objectives

1. Calculate percentages using a percentage circle.
2. Calculate yield percentages.
3. Explain how to conduct yield tests.
4. Calculate as-purchased quantities.
5. Calculate edible-portion quantities.
6. Calculate baker's percentages.
7. Create a formula from a recipe.
8. Explain how ratios are used in the professional kitchen.

Key Terms

- percentage
- yield percentage
- edible-portion (EP) quantity
- as-purchased (AP) quantity
- raw yield test
- cooking-loss yield test
- formula
- baker's percentage
- ratio

CALCULATING PERCENTAGES

Like fractions and decimals, percentages express a part of a whole. A *percentage* is a number that expresses part of a whole in terms of hundredths. One percent is 1 part of 100. The percent symbol (%) is used as an abbreviation for the word percent. For example, if there are 100 jellybeans in a jar and 40 of them are yellow and 60 of them are green, 40 percent (40%) of the jellybeans are yellow and 60 percent (60%) of the jellybeans are green. **See Figure 5-1.**

Florida Tomato Committee

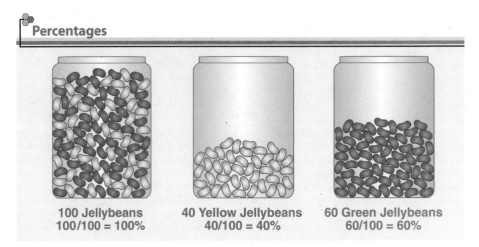

Percentages

100 Jellybeans	40 Yellow Jellybeans	60 Green Jellybeans
100/100 = 100%	40/100 = 40%	60/100 = 60%

Figure 5-1. Percentages are used to express part of a whole in terms of hundredths.

Percentage Guide

In calculations using percentages, the percentage must first be changed to a decimal.

Option 1: Divide the percentage by 100 and drop the percent symbol.

$$40\% = \tfrac{40}{100} = 0.40$$

Option 2: Move the decimal point two places to the left in the percentage and drop the percent symbol.

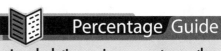

$$40\% = .40 = 0.40$$

To calculate the answer to a problem as a percentage, decimal answers are changed to percentages by multiplying the result by 100 and adding a percent symbol.

$$0.40 \times 100 = 40\%$$

A simple way to understand percentages is through the use of a percentage circle. **See Figure 5-2.** A percentage circle is divided into three sections. The top section of the circle contains the variable P that represents the part of the whole. The lower left section of the circle contains the variable W that represents the whole. The lower right section of the circle contains the variable % that represents the percentage.

To use the percentage circle, any variable that is unknown can be calculated by creating a formula based on where the other two variables are located in the circle. For example, if percentage (%) is the unknown variable, the circle shows the part (P) over the whole (W). This indicates that the percentage can be calculated by dividing the part by the whole.

$$\% = P \div W$$

where
% = percentage
P = part
W = whole

$$percentage = \frac{part}{whole}$$

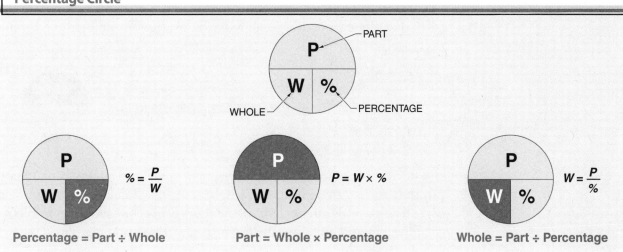

Figure 5-2. The percentage circle is a visual tool used to show how to perform basic percentage calculations.

Even when there are not exactly 100 parts of something, percentages can still be calculated. For example, consider a jar that contains 50 jellybeans. If there were 20 yellow jellybeans and 30 green jellybeans, it would still be true that 40% of the jellybeans are yellow and 60% of the jellybeans are green. This can be demonstrated using the following problem-solving steps and the formula for calculating the percentage.

Problem-Solving Steps

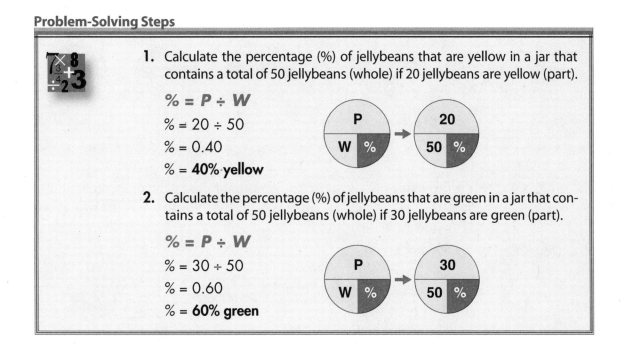

1. Calculate the percentage (%) of jellybeans that are yellow in a jar that contains a total of 50 jellybeans (whole) if 20 jellybeans are yellow (part).

$$\% = P \div W$$
$$\% = 20 \div 50$$
$$\% = 0.40$$
$$\% = \mathbf{40\% \ yellow}$$

2. Calculate the percentage (%) of jellybeans that are green in a jar that contains a total of 50 jellybeans (whole) if 30 jellybeans are green (part).

$$\% = P \div W$$
$$\% = 30 \div 50$$
$$\% = 0.60$$
$$\% = \mathbf{60\% \ green}$$

If the part (P) is the unknown variable, the percentage circle shows the whole (W) and the percentage (%) next to each other, which indicates that the part can be calculated by multiplying the whole by the percentage.

$$P = W \times \%$$

where
P = part
W = whole
% = percentage

part = whole × percentage

The number of jellybeans that are green or yellow can be calculated by multiplying the total number of jellybeans (the whole) by the percentage of each color. This can be demonstrated by using the following problem-solving steps and applying the formula for calculating the part.

Problem-Solving Steps

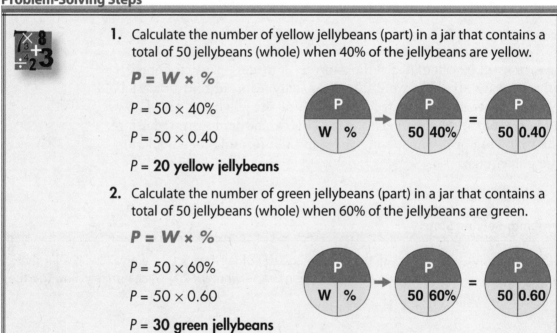

1. Calculate the number of yellow jellybeans (part) in a jar that contains a total of 50 jellybeans (whole) when 40% of the jellybeans are yellow.

$$P = W \times \%$$

$P = 50 \times 40\%$

$P = 50 \times 0.40$

$P = $ **20 yellow jellybeans**

2. Calculate the number of green jellybeans (part) in a jar that contains a total of 50 jellybeans (whole) when 60% of the jellybeans are green.

$$P = W \times \%$$

$P = 50 \times 60\%$

$P = 50 \times 0.60$

$P = $ **30 green jellybeans**

If the whole (W) is the unknown variable, the percentage circle shows the part (P) over the percentage (%). This indicates that the whole can be calculated by dividing the part by the percentage.

$$W = P \div \%$$

where
W = whole
P = part
% = percentage

$$whole = \frac{part}{percentage}$$

The total number of jellybeans (the whole) could be calculated by dividing the part (the number of green or yellow jellybeans) by the percentage of the corresponding color of jellybeans. This can be demonstrated by using the following problem-solving steps and applying the formula for calculating the whole.

Problem-Solving Steps

1. Calculate the total number of jellybeans in a jar (whole) if 40% of the jellybeans are yellow and there are 20 yellow jellybeans (part) in the jar.

   $W = P \div \%$

   $W = 20 \div 40\%$

   $W = 20 \div 0.40$

   $W =$ **50 jellybeans**

2. Calculate the total number of jellybeans in a jar (whole) if 60% of the jellybeans are green and there are 30 green jellybeans (part) in the jar.

   $W = P \div \%$

   $W = 30 \div 60\%$

   $W = 30 \div 0.60$

   $W =$ **50 jellybeans**

Checkpoint 5-1

Master Math™ Applications

1. Define percentage.

   _____

2. If there are 100 peppercorns in a peppercorn grinder and 18 of the peppercorns are green, what percentage of the peppercorns are green?

   _____

3. What is 47% expressed as a decimal?

   _____

4. What is 0.27 expressed as a percentage?

   _____

5. If there are 18 golden delicious apples and 12 granny smith apples in a basket, what percentage are granny smith apples?

_____

_____

6. If there are 50 items on a menu and 20% of the items cost less than $10, how many menu items cost less than $10?

_____

_____

7. If 40% of the cookies on a cookie tray are chocolate chip cookies and there are 20 chocolate chip cookies, how many total cookies are on the tray?

_____

_____

USING YIELD PERCENTAGES

Foodservice operations purchase many products in a form that is different from the way the product will be used. **See Figure 5-3.** For example, meats may need to be trimmed of excess fat and bone, produce may need to be peeled and seeded, or seafood may need to be scaled, skinned, and have the heads and fins removed. In all of these cases, waste is generated.

The terms "as purchased (AP)" and "edible portion (EP)" are commonly used in food service to distinguish between a food item before and after it is trimmed of waste. For example, an *as-purchased (AP) quantity* is the original amount of a food item as it is ordered and received. Likewise, an *edible-portion (EP) quantity* is the amount of a food item that remains after trimming and is ready to be served or used in a recipe. **See Figure 5-4.**

To account for waste, foodservice workers perform calculations based on yield percentages. A *yield percentage* is the edible-portion (EP) quantity of a food item divided by the as-purchased (AP) quantity and is expressed as a percentage. Yield percentages do not apply to food items that are served in the same form as

Trimming Meats

Trimming Excess Fat

Portioning for Use

Figure 5-3. Meats are trimmed of excess fat before being used in a recipe.

they are purchased such as whole pieces of fruit served on a buffet or premade pastries that are simply plated for service.

The three formulas involving yield percentages are used for different reasons.

- **Calculating YP.** A yield percentage is calculated when it cannot be found for a particular food item in any reference material.
- **Calculating AP Quantity.** An AP quantity is calculated when the amount of a food item that is required for a recipe (EP quantity) is known and the amount to be ordered needs to be determined.
- **Calculating EP Quantity.** An EP quantity is calculated when a purchased amount of food (AP quantity) is already on hand and the edible or usable amount of the food needs to be calculated.

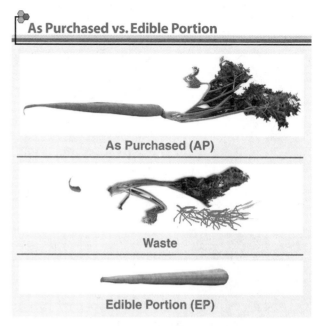

As Purchased (AP)

Waste

Edible Portion (EP)

Figure 5-4. The greens, tips, and peel of whole carrots in AP form are removed and discarded as waste prior to use. The edible portion of the carrot is what remains.

Calculating Yield Percentages

The percentage circle can be modified to help understand how calculations using yield percentages are performed. The circle is modified by first substituting EP quantity for the part (P) since the EP quantity represents the part of the food item that is edible. Then, AP quantity is substituted for the whole (W) since the AP quantity represents the whole amount of the item. Finally, yield percentage (YP) is substituted for the percentage (%). When any two of the three variables are known, the third variable can be calculated. **See Figure 5-5.**

Yield Percentage Circle

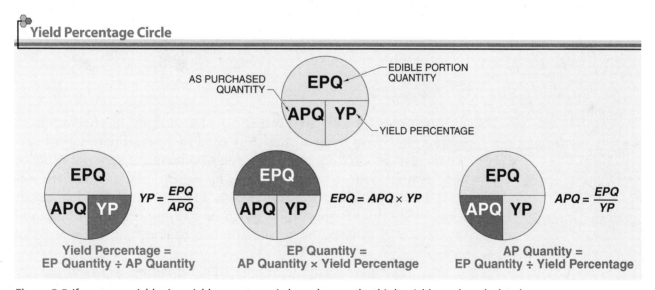

AS PURCHASED QUANTITY

EDIBLE PORTION QUANTITY

EPQ

APQ **YP**

YIELD PERCENTAGE

EPQ

APQ **YP**

$$YP = \frac{EPQ}{APQ}$$

Yield Percentage = EP Quantity ÷ AP Quantity

EPQ

APQ **YP**

$$EPQ = APQ \times YP$$

EP Quantity = AP Quantity × Yield Percentage

EPQ

APQ **YP**

$$APQ = \frac{EPQ}{YP}$$

AP Quantity = EP Quantity ÷ Yield Percentage

Figure 5-5. If any two variables in a yield percentage circle are known, the third variable can be calculated.

Common yield percentages are available in various reference tables. **See Appendix.** However, if a yield percentage for a particular food item is not known and cannot be found in reference material, it can be calculated by performing a raw yield test or a cooking-loss yield test.

A *raw yield test* is a procedure used to determine the yield percentage of a food item that is trimmed of waste prior to being used in a recipe. Yield percentage is calculated by using the following formula:

YP = EPQ ÷ APQ

where
YP = yield percentage
EPQ = EP quantity
APQ = AP quantity

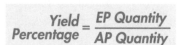

$$\text{Yield Percentage} = \frac{\text{EP Quantity}}{\text{AP Quantity}}$$

To perform the raw yield test, a food item is purchased and weighed to determine the AP quantity. The food item is then trimmed and the edible portion is weighed to determine the EP quantity. For example, to calculate the yield percentage of carrots, the AP quantity of carrots is weighed. Then, the carrots are peeled and the greens and tips are removed as waste (unless the trimmings can be used in another recipe such as a stock). The EP quantity is determined by weighing the cleaned carrots. If 10 pounds of carrots weigh 8.5 pounds after being trimmed, the yield percentage can be calculated using the formula as follows:

YP = EPQ ÷ APQ

YP = 8.5 lb ÷ 10 lb

YP = 0.85

YP = **85%**

The other type of yield test is called a cooking-loss yield test. A *cooking-loss yield test* is a procedure used to determine the yield percentage of a food item that loses weight during the cooking process. For example, meat loses weight as fat is rendered and moisture is lost during cooking. **See Figure 5-6.** The cooking-loss yield test is used when the EP quantity is based on the amount of cooked food to be served as opposed to the amount of raw food to be used in a recipe.

The yield percentage is calculated by dividing the EP quantity by the AP quantity. To perform the test, a food item is weighed to determine the AP quantity. The food item is then weighed again after being trimmed and/or cooked to determine the EP quantity. For example, a hamburger that weighs 8 ounces prior to cooking might weigh 6 ounces after cooking. The yield percentage in this case can be calculated by using the formula as follows:

YP = EPQ ÷ APQ

$YP = 6 \div 8$

$YP = 0.75$

$YP = \textbf{75\%}$

Cooking-Loss Yield Test

Hamburger Before Cooking   Hamburger After Cooking

Figure 5-6. Meat can lose up to 30% of its weight as moisture is lost and fat is rendered during roasting.

Calculating As-Purchased Quantities

Calculating an AP quantity is done to determine the proper amount of an ingredient to order to make a given quantity of food. The AP quantity is calculated by using the following formula:

APQ = EPQ ÷ YP

where
APQ = AP quantity
EPQ = EP quantity
YP = yield percentage

$$\frac{AP}{Quantity} = \frac{EP\ Quantity}{Yield\ Percentage}$$

Consider a banquet for 100 people that includes grilled salmon on the menu. The recipe for the grilled-salmon dish calls for a 0.5-pound fillet of salmon per serving. The total EP quantity of salmon can be calculated by multiplying the amount of salmon required per person by the number of people attending the banquet.

$$100\ \cancel{\text{people}} \times \frac{0.5\ \text{lb}}{\cancel{\text{person}}} = 50\ \text{lb}$$

If the chef decides to order whole salmon for the banquet, the heads, fins, skin, and bones of the salmon will need to be removed. **See Figure 5-7.** If the yield percentage for cutting salmon fillets from whole salmon is 50%, the chef can then calculate how much whole salmon to order. The quantity of salmon to be ordered can be calculated by applying the formula for AP quantity as follows:

APQ = EPQ ÷ YP

$APQ = 50\ \text{lb} \div 50\%$

$APQ = 50\ \text{lb} \div 0.50$

$APQ = \textbf{100 lb}$

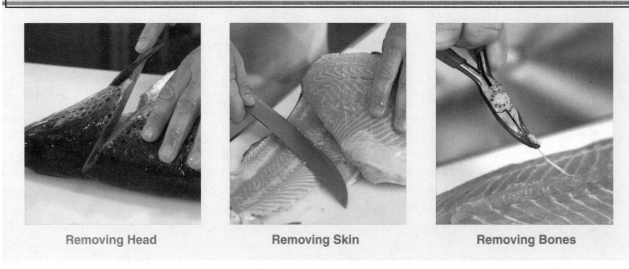

| Removing Head | Removing Skin | Removing Bones |

Figure 5-7. The heads, skin, and bones of whole fish are removed to produce boneless-skinless fillets.

Calculating Edible-Portion Quantities

Calculating an EP quantity is done to determine how much food can be prepared using an amount of an ingredient that is already on hand. The EP quantity can be calculated using the following formula:

EPQ = APQ × YP

where

EPQ = EP quantity
APQ = AP quantity
YP = yield percentage

*EP Quantity = AP Quantity ×
Yield Percentage*

Sometimes it is necessary to calculate how much food can be prepared from an amount of food that is already on hand. For example, a chef decides to take advantage of a special price being offered by the meat supplier and purchases 40 pounds of whole beef tenderloin. The chef creates a special recipe for beef stroganoff and needs to know how many servings of the recipe can be made using the beef tenderloin.

The recipe calls for 6 ounces per serving of fully trimmed beef tenderloin cut into 1-inch pieces. The chef knows from experience that after trimming the fat and connective tissue from the beef tenderloin only 75% (YP) of the tenderloin will be used. The number of servings of beef stroganoff that can be prepared is calculated using the formula for EP quantity and applying the following problem-solving steps.

The Beef Checkoff

1. Calculate the edible-portion quantity of beef tenderloin by multiplying the as-purchased quantity by the yield percentage.

EPQ = APQ × YP

EPQ = 40 lb × 75%

EPQ = 40 lb × 0.75

EPQ = **30 lb**

EPQ	EPQ	EPQ
APQ | YP	40 | 75%	40 | 0.75

→ =

2. Calculate the number of servings of beef stroganoff that can be made by using the 30 pounds of trimmed beef tenderloin.

$$30\ \text{lb} \times \frac{16\ \text{oz}}{1\ \text{lb}} \times \frac{1\ \text{serving}}{6\ \text{oz}} = \frac{480\ \text{servings}}{6} = \textbf{80 servings}$$

Factors Affecting Yield Percentages

Calculations involving yield percentages are performed regularly in the professional kitchen with good results. However, special attention should be given to how the results for yield percentage calculations are rounded. **See Figure 5-8.**

Rounding Yield Percentage Calculation Results

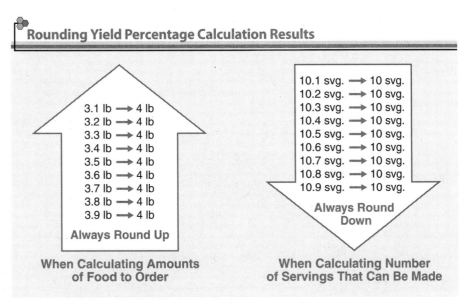

Figure 5-8. Special rules apply to how the results for yield percentage calculations are rounded depending on the quantity being calculated.

For example, if it is determined that 30.4 pounds of fish should be ordered for a banquet, this decimal result would be rounded down to 30 pounds using standard math rules. However, rounding down would result in the banquet coming up short on the amount of fish needed.

For yield percentage calculations related to ordering food, the results should always be rounded up to ensure that enough food is purchased. However, if a yield percentage calculation indicates that 14.8 portions can be made from a given recipe, the result should be rounded down to 14 because the recipe will not make 15 full servings.

Employee skill levels can also dramatically affect yield percentages. Butchering meats is a skill that requires a lot of practice to master. Therefore, a trained butcher will experience higher yield percentages when trimming meats than an entry-level cook assigned to do the same task.

Another factor affecting yield percentage is the condition and size of the initial product. For example, fruits that are unblemished will have higher yield percentages than bruised fruits that require the removal of undesirable parts. Large potatoes have a higher yield percentage than small potatoes. More waste is generated by peeling many small potatoes compared to peeling fewer large potatoes.

Checkpoint 5-2

1. Define yield percentage.

   _____

   _____

2. Define edible-portion (EP) quantity.

   _____

   _____

3. Define as-purchased (AP) quantity.

   _____

   _____

4. Describe the difference between a raw yield test and a cooking-loss yield test.

   _____

   _____

5. What was the yield percentage for beets if, in a raw yield test, the AP quantity of the beets was 2 pounds and 0.5 pounds of waste was generated trimming the beets?

   _____

   _____

Checkpoint 5-2 (continued)

6. If a recipe calls for 8 pounds of peeled potatoes, how many pounds of unpeeled potatoes are required if the yield percentage is 80%?

_____

_____

7. A recipe for grilled pork tenderloin calls for 6 ounces of pork tenderloin per serving and there are 15 pounds of pork tenderloin available. How many servings can be made if the yield percentage for pork tenderloin is 85%?

_____

_____

USING FORMULAS AND BAKER'S PERCENTAGES

The accurate measurement of ingredients in the bakeshop is especially critical. In other types of cooking, the flavor of a dish can be adjusted during the cooking process. Baking is much different from other cooking techniques because once a product is prepared and placed in the oven to bake, there is no opportunity to make any changes. Because of this, formulas are used instead of recipes in the bakeshop. A *formula* is a recipe format in which all ingredient quantities are provided as baker's percentages. A *baker's percentage* is the weight of a particular ingredient expressed as a percentage based on the weight of the main ingredient in a formula. **See Figure 5-9.** Formulas help to ensure that measurements are performed consistently and accurately.

If flour is the main ingredient in a formula, its baker's percentage is set to 100% and all other ingredients in the formula are expressed as a percentage based on the weight of flour. For example, if the formula also uses sugar and the baker's percentage of sugar in the formula is 50%, it means that no matter how much flour is used to prepare that formula, the weight of sugar used will always be equal to 50% of the weight of flour. It does *not* mean that 50% of the total formula is made up of sugar. In fact, the sum of the baker's percentages in a formula will always be greater than 100%.

Formula for Baguettes

Baguettes

Ingredients	Baker's Percentage
Bread Flour (main ingredient)	100%
Water	75%
Fresh Yeast	2%
Salt	1%
Total:	**178%**

Procedure:

1. Place flour, water, and yeast in mixing bowl and mix slowly until combined.
2. Add salt and increase mixer speed to medium. Continue kneading for 10 minutes or until a smooth dough forms.
3. Transfer dough to a lightly greased bowl and allow to proof until doubled in size.
4. Punch down the dough and allow to rest for a few minutes.
5. Divide the dough into 10-ounce balls and form into baguettes 20 inches long.
6. Proof baguettes until double in size and score each before baking.
7. Bake at 450°F for approximately 25 to 30 minutes.

Figure 5-9. The ingredient quantities in formulas are expressed as baker's percentages instead of volume or weight units.

If a formula uses two types of flour, the two types of flour combined are considered the main ingredient and will have a total baker's percentage of 100%. For example, a formula for rye bread may contain white flour with a baker's percentage of 75% and rye flour with a baker's percentage of 25%. In this case, the two types of flour will have a total baker's percentage of 100% (75% + 25% = 100%).

Calculating with Baker's Percentages

Two important rules must be followed when using formulas with baker's percentages. First, all of the items must be weighed (not measured by volume). Second, all of the measurements must have the same unit of measure.

The first step when calculating the weights of the ingredients using a formula is to determine the weight of the main ingredient. Sometimes, the weight of the main ingredient is stated directly. For example, a baker may decide to make a batch of French bread using 20 pounds of bread flour. In other cases, the weight of the main ingredient may need to be calculated based on the desired yield of the formula. When a specific total yield is desired, the weight of the main ingredient can be calculated using the following formula:

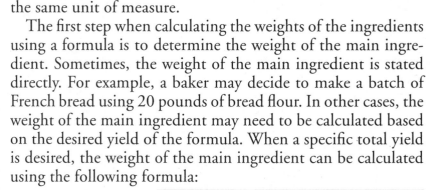

$$WM = DY \div TBP$$

$$\text{Weight of Main Ingredient} = \frac{\text{Desired Yield}}{\text{Total Baker's Percentage}}$$

where

WM = weight of main ingredient
DY = desired yield
TBP = total baker's percentage

For example, if the total of the baker's percentages in a hard-roll dough recipe is 182%, and the desired yield is 30 pounds of dough, the weight of the main ingredient (bread flour) is 16.5 pounds.

$WM = DY \div TBP$
$WM = 30 \text{ lb} \div 182\%$
$WM = 30 \text{ lb} \div 1.82$
$WM = 16.484 = \textbf{16.5 lb}$

Once the weight of the main ingredient is determined, the remaining ingredient weights are calculated. This is done by multiplying the baker's percentage for each of the remaining ingredients in the formula by the weight of the main ingredient. **See Figure 5-10.** The weight of the remaining ingredients are calculated using the following formula:

$$WI = WM \times BP$$

$$\text{Weight of} \atop \text{Ingredient} = {\text{Weight of} \atop \text{Main Ingredient}} \times {\text{Baker's} \atop \text{Percentage}}$$

where
WI = weight of ingredient
WM = weight of main ingredient
BP = baker's percentage

Using Formulas

Hard-Roll Dough (30-lb Batch)		
Ingredients	Baker's Percentage	Ingredient Weight* ($WI = WM \times BP$)
Bread Flour (main ingredient)	100%	16.5 lb × 100% = 16.5 lb × 1.00 = **16.5 lb**
Salt	3%	16.5 lb × 3% = 16.5 lb × 0.03 = **0.5 lb**
Sugar	5%	16.5 lb × 5% = 16.5 lb × 0.05 = **0.8 lb**
Shortening	6%	16.5 lb × 6% = 16.5 lb × 0.06 = **1.0 lb**
Egg Whites	3%	16.5 lb × 3% = 16.5 lb × 0.03 = **0.5 lb**
Water	60%	16.5 lb × 60% = 16.5 lb × 0.60 = **9.9 lb**
Yeast	5%	16.5 lb × 5% = 16.5 lb × 0.05 = **0.8 lb**
Total	**182%**	**30 lb**

* Results rounded to the tenths place.

Figure 5-10. When using a formula, ingredient weights are calculated by multiplying the baker's percentage for each ingredient by the amount of the main ingredient to be used.

For example, in the hard-roll dough formula, bread flour is the main ingredient. If 16.5 pounds of bread flour is used to make the dough and the baker's percentage for salt in this formula is 3%, the weight of salt required is calculated as follows:

$$WI = WM \times BP$$

$WI = 16.5 \text{ lb} \times 3\%$
$WI = 16.5 \text{ lb} \times 0.03$
$WI = \textbf{0.5 lb}$

Flour is not always the main ingredient in a formula. In fact, some formulas will not contain any flour. For example, dried apricots are the main ingredient in a recipe for an apricot filling. If the baker's percentage for water is 20% and the filling is to be made using 10 pounds of dried apricots, the weight of water required is calculated as follows:

$$WI = WM \times BP$$

$WI = 10 \text{ lb} \times 20\%$
$WI = 10 \text{ lb} \times 0.20$
$WI = \textbf{2.0 lb}$

Creating Formulas from Recipes

Sometimes recipes that have been written in a traditional format, without a formula provided, need to be modified for use in the professional kitchen. To create a formula from a recipe, the first step is to identify the main ingredient. The next step is to convert all of the ingredient amounts in the recipe to the same weight units of measure. Then, the weight of each ingredient is divided by the weight of the main ingredient. The result, in decimal form, is changed to a percentage. A baker's percentage is calculated by using the following formula:

$$BP = WI \div WM$$

where

BP = baker's percentage

WI = weight of ingredient

WM = weight of main ingredient

$$\text{Baker's Percentage} = \frac{\text{Weight of Ingredient}}{\text{Weight of Main Ingredient}}$$

For example, a white cake recipe may call for 5 pounds of cake flour, 3.5 pounds of shortening, 5 ounces of baking powder, and a number of other ingredients. To create a formula from this recipe, all of the original ingredient amounts need to be converted to a common unit of measure. First, the amount of the main ingredient, cake flour, is converted from 5 pounds to 80 ounces.

$$5 \, \text{lb} \times \frac{16 \, \text{oz}}{1 \, \text{lb}} = 80 \, \text{oz}$$

To calculate the baker's percentage of shortening, the weight of shortening called for in the recipe is first converted to ounces.

$$3.5 \, \text{lb} \times \frac{16 \, \text{oz}}{1 \, \text{lb}} = 56 \, \text{oz}$$

Then the formula for calculating a baker's percentage is used.

$$BP = WI \div WM$$

BP = 56 oz ÷ 80 oz

BP = 0.70

BP = **70%**

The process for creating a formula is repeated for every ingredient in the recipe. **See Figure 5-11.** In this example, conversion calculations are not necessary for the ingredient measurements provided in ounces.

Creating Formulas from Recipes

Ingredients	Ingredient Weight (WI)	Ingredient Weight Converted to Ounces	Baker's Percentage (BP = WI ÷ WM)
		White Cake	
Cake Flour (main ingredient)	**5 lb (WM)**	5 lb × 16 oz/lb = **80 oz**	80 oz ÷ 80 oz = 1.0 = **100%**
Shortening	3.5 lb	3.5 lb × 16 oz/lb = **56 oz**	56 oz ÷ 80 oz = 0.7 = **70%**
Sugar	6.25 lb	6.25 lb × 16 oz/lb = **100 oz**	100 oz ÷ 80 oz = 1.25 = **125%**
Salt	3 oz	**3 oz**	3 oz ÷ 80 oz = 0.0375 = **3.75%**
Baking Powder	5 oz	**5 oz**	5 oz ÷ 80 oz = 0.0625 = **6.25%**
Water	1.75 lb	1.75 lb × 16 oz/lb = **28 oz**	28 oz ÷ 80 oz = 0.35 = **35%**
Nonfat Dry Milk	5 oz	**5 oz**	5 oz ÷ 80 oz = 0.0625 = **6.25%**
Eggs	1.25 lb	1.25 lb × 16 oz/lb = **20 oz**	20 oz ÷ 80 oz = 0.25 = **25%**
Egg Whites	2 lb	2 lb × 16 oz/lb = **32 oz**	32 oz ÷ 80 oz = 0.4 = **40%**
Water	2 lb	2 lb × 16 oz/lb = **32 oz**	32 oz ÷ 80 oz = 0.4 = **40%**
Vanilla Extract	2 oz	**2 oz**	2 oz ÷ 80 oz = 0.025 = **2.5%**

Figure 5-11. A recipe can be converted into a formula after converting the recipe ingredient quantities to a common weight unit of measure.

Checkpoint 5-3

1. Define formula.

_____

2. Define baker's percentage.

_____

_____

3. What percentage is assigned to the main ingredient in a formula?

_____

_____

4. If the main ingredient in a formula is sugar and the percentage for vanilla extract in the recipe is 2%, how much vanilla extract is needed if 8 pounds of sugar are used?

_____

_____

5. If a biscuit recipe calls for 2.5 pounds of flour (the main ingredient) and 1.5 pounds of butter, what would the baker's percentage be for the butter if the recipe is converted to a formula?

CALCULATING RATIOS

Some recipes are written as ratios. A *ratio* is a mathematical way to represent the relationship between two or more numbers or quantities. In the professional kitchen, a ratio is expressed in terms of parts and can be shown in different ways. For example, a ratio of "three parts to one part" can be expressed using numbers with the word "to" between them (3 to 1) or with a colon between them (3:1).

Consider a vinaigrette that is made with a ratio of 3 parts oil to 1 part vinegar by volume. The term "parts" is used as a generic term because it does not matter what volume unit of measure is used when making the vinaigrette as long as the unit is the same for both ingredients. For example, a vinaigrette will taste the same if made with 3 cups of oil and 1 cup of vinegar or 3 quarts of oil and 1 quart of vinegar.

Ratios can also be converted to fractions and percentages. In the case of the vinaigrette, the ratio consists of a total of four parts (3 parts oil + 1 part vinegar). Since the oil represents three parts out of four and the vinegar represents one part out of four, a vinaigrette with this ratio can be thought of as containing ¾ oil and ¼ vinegar. By converting the fractions ¾ and ¼ to percentages, the vinaigrette can also be thought of as containing 75% oil and 25% vinegar. **See Figure 5-12.**

Forms and Tables

Converting Ratios to Fractions and Percentages

Vinaigrette				
	Ratio	Total Parts	Fractions	Percentages
3 PARTS OIL / 1 PART VINEGAR	3:1	3 + 1 = 4	¾ Oil	3 ÷ 4 = 0.75 / 75% Oil
			¼ Vinegar	1 ÷ 4 = 0.25 / 25% Oil

Mirepoix				
	Ratio	Total Parts	Fractions	Percentages
1 PART CELERY / 2 PARTS ONION / 1 PART CARROTS	2:1:1	2 + 1 + 1 = 4	2/4 = ½ Onions	1 ÷ 2 = 0.50 / 50% Onions
			¼ Carrots	1 ÷ 4 = 0.25 / 25% Carrots
			¼ Celery	1 ÷ 4 = 0.25 / 25% Celery

Figure 5-12. Ratios can be converted into fractions and percentages.

An example of a ratio with three ingredients is the 2:1:1 ratio for mirepoix (a mixture of 2 parts onion, 1 part carrot, and 1 part celery used when making stock). The ratio for mirepoix contains a total of four parts where onions represent two parts of four, and carrots and celery each represent one part of four. Therefore, mirepoix is made of 50% onions, and 25% each of celery and carrots.

Checkpoint 5-4

1. Define ratio.

_____

2. If the ratio of sugar to shortening in a cookie recipe is 2:1, what fraction of the recipe is sugar?

_____

_____

3. If the ratio of flour to sugar in a cake recipe is 3:2, what percent of the recipe is flour?

_____

_____

USING RATIOS

A ratio is similar to a baker's formula in that the ratio does not contain actual measurements for the ingredients. Instead, a ratio describes how the ingredients are used in relation to one another. The simplest ratio used in the professional kitchen is a ratio of 1:1 (also referred to as equal parts).

Roux (a mixture of fat and flour used to thicken soups and sauces) is based on a 1:1 ratio by weight of fat to flour. This simply means that 1 pound of flour will be needed for every 1 pound of fat used to make the roux. In a ratio of 1:1 there are two total parts. By converting the ratio to fractions, the recipe for roux becomes ½ flour and ½ fat.

To calculate the amount of each ingredient required to make a given amount of roux, multiply the total desired yield by the fraction associated with each ingredient. With a ratio of 1:1, it is easy to calculate that 10 pounds of roux would be made from 5 pounds of fat and 5 pounds of flour.

$$\text{Fat: } 10 \text{ lb} \times \frac{1}{2} = \frac{10 \text{ lb}}{2} = \textbf{5 lb}$$

$$\text{Flour: } 10 \text{ lb} \times \frac{1}{2} = \frac{10 \text{ lb}}{2} = \textbf{5 lb}$$

For some ratios, it is important to know whether to use weight units or volume units when calculating ingredient quantities. For example, the ratio for roux is 1:1 flour to fat "by weight." The ratio does not work using volume units. The roux will turn out correctly whether made with 1 ounce of flour and 1 ounce of fat or 1 pound of flour and 1 pound of fat. However, it will not turn out correctly if made with 1 cup of flour and 1 cup of fat because 1 cup of flour does not weigh the same as 1 cup of fat.

When a ratio must be used in terms of weight only or volume only, the ratio will be followed by the words "by weight" or "by volume." Consider a ratio for pie dough of three parts flour to two parts butter to one part water (3:2:1) by weight. If the desired yield of pie dough is 12 pounds, how much of each ingredient would be required?

The first step would be to convert the ratio to fractions by dividing each number in the ratio by the total number of parts (3 + 2 + 1 = 6). The fraction for flour would be $\frac{3}{6}$ (which reduces to $\frac{1}{2}$), the fraction for butter would be $\frac{2}{6}$ (which reduces to $\frac{1}{3}$), and the fraction for water would be $\frac{1}{6}$. By multiplying each of the fractions by the total desired amount of dough (12 pounds), it can be calculated that the recipe would require 6 pounds of flour ($\frac{1}{2} \times 12 = 6$), 4 pounds of butter ($\frac{1}{3} \times 12 = 4$), and 2 pounds of water ($\frac{1}{6} \times 12 = 2$). **See Figure 5-13.**

Media Clips Using Ratios

Using Pie Dough Ratios

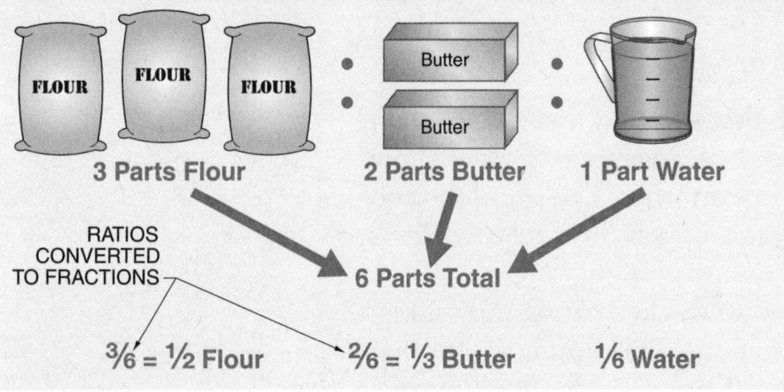

DESIRED RECIPE YIELD

12 Pounds Pie Dough	
Ingredient	*Quantity Required*
Flour	$\frac{1}{2} \times 12$ lb = 12 lb/2 = 6 lb
Butter	$\frac{1}{3} \times 12$ lb = 12 lb/3 = 4 lb
Water	$\frac{1}{6} \times 12$ lb = 12 lb/6 = 2 lb

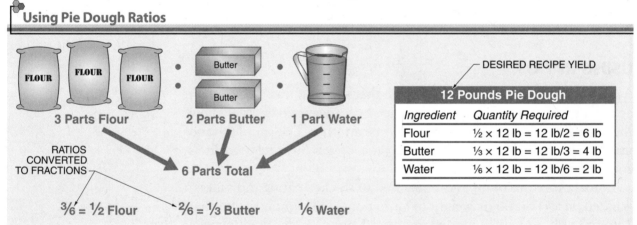

Figure 5-13. When using a ratio, ingredient quantities can be calculated by multiplying the fractions of the ratio by the desired yield of the recipe.

In some cases it may be necessary to calculate the total amount of a recipe that can be made based on the amount of one ingredient. For example, cooked white rice is made with a ratio of 2 parts liquid to 1 part uncooked white rice by volume. If the rice is to be prepared using chicken stock and there are only 6 quarts of chicken stock available, how much total rice can be prepared?

In this case, the ratio has three total parts (2 + 1 = 3). The liquid (chicken stock) is $\frac{2}{3}$ of the ratio and the uncooked rice is $\frac{1}{3}$ of the ratio. To calculate the total amount of cooked rice that can be prepared, the amount of chicken stock is divided by the fraction of the recipe that is chicken stock.

6 qt ÷ ⅔ = 6 qt × ³⁄₂ = 18 qt/2 = **9 qt**

Then, to calculate how much uncooked rice is required, the total amount of cooked rice is multiplied by the fraction of the recipe that is uncooked rice.

9 qt × ⅓ = 9 qt/3 = **3 qt**

Therefore, 9 quarts of cooked rice could be made from 6 quarts of chicken stock and 3 quarts of uncooked rice.

Checkpoint 5-5

Master Math™ Applications

1. Cookie dough is made with a ratio of 3 parts flour to 2 parts sugar to 1 part butter. If 9 pounds of cookie dough needs to be made, how many pounds of sugar are required?

   _____

2. A creamy salad dressing is made with a ratio of 4 parts mayonnaise to 1 part sour cream. If there are 2 quarts of sour cream available, how many quarts of salad dressing can be made?

   _____

Chapter 5 Summary

Quick Quiz® Chapter 5

Flash Cards

Percentages are a way of expressing the part of a whole in terms of hundredths. The variables in a percentage calculation are the part, the whole, and the percentage. If any two of the variables are known, the third can be calculated.

Yield percentages (YP) are used to define how much of a food item, in an as-purchased (AP) form, is edible or usable in a recipe. In a yield percentage calculation the AP quantity of a food item is the whole and the edible-portion (EP) quantity is the part. A yield percentage can be found by performing a raw yield test or a cooking-loss yield test. The results of yield test calculations are used to ensure that the proper amount of food is ordered. Results are also used to calculate how much edible or usable food can be produced from an AP amount of food.

Formulas are used in the bakeshop instead of recipes and are written using baker's percentages. These percentages represent ingredient quantities in terms of percentages based on the amount of main ingredient in the formula. In addition to recipes and formulas, ratios are used in the professional kitchen. Like formulas, ratios do not contain actual ingredient amounts but provide an indication of how ingredients are used in proportion to one another. Both formulas and ratios provide simple ways to calculate ingredient quantities based on a desired yield.

Checkpoint 5-1

1. A *percentage* is a number that expresses part of a whole in terms of hundredths.
2. 18%   (18 ÷ 100 = 0.18 = 18%)
3. 0.47
4. 27%
5. 40%   (18 + 12 = 30 and 12 ÷ 30 = 0.40 = 40%)
6. 10   (50 × 20% = 50 × 0.20 = 10)
7. 50   (20 ÷ 40% = 20 ÷ 0.40 = 50)

Checkpoint 5-2

1. *Yield percentage* is the percentage of a food item in as-purchased form that is edible or usable in a recipe.
2. *Edible-portion (EP) quantity* is the amount of a food item that can be used in a recipe after unusable parts are trimmed away.
3. *As-purchased (AP) quantity* is the original amount of a food item as ordered and received.
4. A *raw yield test* provides the yield percentage of a food item before the item is used or cooked in a recipe and a *cooking-loss yield test* provides the yield percentage for a food item that loses weight during cooking.
5. 75%   (2 lb – 0.5 lb = 1.5 lb and 1.5 lb ÷ 2.0 lb = 0.75 = 75%)
6. 10 lb   (8 lb ÷ 80% = 8 lb ÷ 0.80 = 10 lb)
7. 34 servings   (15 × 16 oz = 240 oz, 240 oz × 85% = 240 oz × 0.85 = 204 oz, and 204 × 1 serving/6 = 34 servings)

Checkpoint 5-3

1. A *formula* is a recipe format in which all of the ingredient quantities are provided as baker's percentages.
2. A *baker's percentage* is the amount of a particular ingredient expressed as a percentage of the main ingredient in a baker's formula.
3. 100%
4. 0.16 lb   (8 lb × 2% = 8 lb × .02 = 0.16 lb)
5. 60%   (1.5 lb ÷ 2.5 lb = 0.6 = 60%)

Checkpoint 5-4

1. A *ratio* is a mathematical way to represent the relationship between two or more numbers or quantities.
2. ⅔
3. 60%   (3 + 2 = 5 and ⅗ = 0.60 = 60%)

Checkpoint 5-5

1. 3 lb   (3 + 2 + 1 = 6, ²⁄₆ = ⅓, and ⅓ × 9 lb = 3 lb)
2. 10 qt   (4 + 1 = 5, and 2 qt ÷ ⅕ = 10 qt)

Calculating Food Costs and Menu Prices

Many costs must be monitored and managed in order to run a successful foodservice operation. Among the most significant costs of a foodservice operation are food and beverage costs. In order to calculate appropriate menu prices, the cost to prepare the menu items must be known. It is also important to know how food and beverage costs are used to help calculate menu prices and how customer perceptions influence menu pricing.

Chapter Objectives

1. Identify as-purchased costs on invoices.
2. Calculate as-purchased unit costs.
3. Calculate edible-portion unit costs.
4. Calculate as-served costs of menu items.
5. Calculate food cost percentages of menu items.
6. Calculate overall food cost percentages of foodservice operations.
7. Calculate menu prices using the food cost percentage pricing method.
8. Explain perceived value pricing.
9. Calculate menu prices based on the contribution margin pricing method.
10. Use pricing forms to help establish menu prices.

Key Terms

- as-purchased (AP) cost
- invoice
- unit cost
- as-purchased (AP) unit cost
- edible-portion (EP) unit cost
- as-served (AS) cost
- food cost percentage
- menu-item food cost percentage
- overall food cost percentage
- target food cost percentage
- beverage cost percentage
- target price
- perceived value pricing
- contribution margin
- pricing form

IDENTIFYING AS-PURCHASED COSTS

The most fundamental food and beverage cost is referred to as the as-purchased (AP) cost. An *as-purchased (AP) cost* is the amount paid for a product in the form it was ordered and received. The AP costs for food and beverage products are documented on invoices. An *invoice* is a document provided by a supplier that lists the items delivered to a foodservice operation and the prices of those items.

Foodservice operations typically buy food products from suppliers in bulk quantities. Examples of bulk quantities include cases of hamburger patties, tubs of ice cream, crates of milk, or pails of pickles. If a 20-pound case of 80 hamburger patties costs $30.00, the AP cost is $30.00 for the entire case. Likewise, if the invoice price of a 3-gallon tub of ice cream is $12.00, the AP cost for the tub of ice cream is $12.00. **See Figure 6-1.**

As-Purchased Costs

DATE	PURCHASE ORDER NO. RA 05353066	PAGE NO. 1 of 1	DELIVERY DATE

SHIP TO:	Acme Foods Silver Spring Rd Stony Branch, OH 66005	**VENDOR** Stone Cold Foods

CHARGE & INVOICE TO:	Acme Foods Silver Spring Rd Stony Branch, OH 66005

QUANTITY	UNIT	ITEM ID — DESCRIPTION	UNIT PRICE	ITEM TOTAL
1	Case (80 ct)	Frozen Hamburger Patties (20 lb)	$30.00	$30.00
1	1 Tub (3 gal.)	Signature Ice Cream	$12.00	$12.00

I certify that sufficient funds are available for this purchase.

_____
Signature
Purchasing Manager
Title

_____
Date

Tax (if applicable)	
Shipping	$6.00
Total	

Type of Payment
☐ Payment Enclosed
☑ COD
☐ Credit

Figure 6-1. As-purchased (AP) costs are listed on invoices from suppliers.

Beverages are also typically purchased in bulk quantities such as 24-bottle cases of iced tea, 10-bottle cases of wine, or half barrels (kegs) of beer. As with food products, the AP costs of beverage products are equal to the prices listed on the invoices from beverage suppliers.

1. Define as-purchased (AP) cost.

2. Where can an AP cost be found?

3. What is the AP cost of a 25-pound bag of rice that sells for $15 per bag?

4. What is the AP cost of a case of 12 bottles of mineral water that sells for $18 per case?

CALCULATING UNIT COSTS

When a product is purchased in bulk, it is often necessary to calculate a unit cost based on the AP cost. A *unit cost* is the cost of a product per unit of measure. Unit costs can be based on weight (ground turkey @ $2.75 per pound), volume (milk @ $3.00 per gallon), or count (bread @ $2.00 per loaf).

Calculating As-Purchased Unit Costs

Unit costs are initially calculated based on AP costs. The *as-purchased (AP) unit cost* is the unit cost of a food item based on the form in which it is ordered and received. The AP unit cost of an item is calculated by applying the following formula:

APU = APC ÷ NU

where
APU = AP unit cost
APC = AP cost
NU = number of units

$$AP\ Unit\ Cost = \frac{AP\ Cost}{Number\ of\ Units}$$

For example, to calculate the AP unit cost of an egg, divide the AP cost for a case of eggs by the number of eggs in the case. If the price paid for a case of eggs is $14.40 and there are 15 dozen eggs in the case, the AP unit cost would be $0.08 per egg.

APU = APC ÷ NU

APU = $14.40 ÷ 15 dozen

APU = $14.40 ÷ (15 dozen × 12 eggs/dozen)

APU = $14.40 ÷ 180 eggs

APU = **$0.08/egg**

Media Clips · Calculating AP Unit Costs

The unit of measure used to calculate a unit cost should be based on how the product is used in recipes. Sometimes it may be helpful to calculate the unit cost of an item in more than one unit of measure. For example, some recipes may call for fluid ounces of heavy cream and other recipes may call for quarts of heavy cream. **See Figure 6-2.**

Calculating As-Purchased Unit Costs

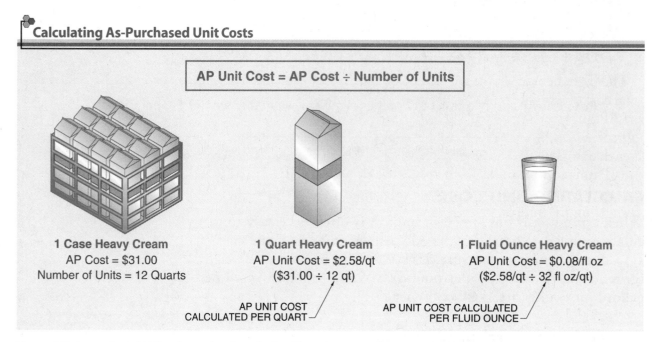

AP Unit Cost = AP Cost ÷ Number of Units

1 Case Heavy Cream
AP Cost = $31.00
Number of Units = 12 Quarts

1 Quart Heavy Cream
AP Unit Cost = $2.58/qt
($31.00 ÷ 12 qt)

AP UNIT COST CALCULATED PER QUART

1 Fluid Ounce Heavy Cream
AP Unit Cost = $0.08/fl oz
($2.58/qt ÷ 32 fl oz/qt)

AP UNIT COST CALCULATED PER FLUID OUNCE

Figure 6-2. As-purchased (AP) unit costs can be calculated based on more than one unit of measure.

Calculating Edible-Portion Unit Costs

A similar calculation is performed to determine the edible-portion (EP) unit cost based on the AP unit cost. The *edible-portion (EP) unit cost* is the unit cost of a food or beverage item after taking into account the cost of the waste generated by trimming. Unless a food item has a yield percentage of 100%, the EP unit cost will always be higher than the AP unit cost. Consider a mashed potato recipe that calls for 10 pounds of peeled potatoes. The cost of the potatoes for the recipe will be more than the cost of 10 pounds of whole potatoes because more than 10 pounds of whole potatoes will need to be peeled in order to yield 10 pounds of peeled potatoes.

The EP unit cost of a food item is calculated by applying the following formula:

$$EPU = APU \div YP$$

where
EPU = EP unit cost
APU = AP unit cost
YP = yield percentage

$$EP\ Unit\ Cost = \frac{AP\ Unit\ Cost}{Yield\ Percentage}$$

For example, if a recipe calls for peeled potatoes and the AP unit cost of whole potatoes is $0.50 per pound and the yield percentage is 80%, what is the EP unit cost of the peeled potatoes?

EPU = APU ÷ YP

EPU = $0.50/lb ÷ 80%

EPU = $0.50/lb ÷ 0.80

EPU = $0.625/lb = **$0.63/lb**

The ingredient amounts provided in a recipe are EP quantities. The total cost of an ingredient in a recipe can be calculated by multiplying the EP quantities provided in the recipe by the EP unit cost for each ingredient. For example, what is the total cost of onions in a recipe for French onion soup that calls for 3 pounds of peeled and sliced onions if the AP unit cost of whole onions is $0.40 per pound and the yield percentage is 93%? The answer can be found by applying the formula for calculating EP unit cost and following these problem-solving steps.

Problem-Solving Steps

1. Calculate the edible-portion unit cost for whole onions.

   EPU = APU ÷ YP

   EPU = $0.40/lb ÷ 93%

   EPU = $0.40/lb ÷ 0.93

   EPU = **$0.43/lb**

2. Calculate the total cost of onions in the recipe by multiplying the edible-portion quantity by the edible-portion unit cost.

   3 lb × $0.43/lb = **$1.29**

When calculating the EP unit cost of an ingredient, the unit of measure in the EP quantity and the AP unit cost must be the same. For example, if the EP quantity of sugar in a cake recipe is 12 ounces and the AP unit cost of sugar is $0.64 per pound, the AP unit cost will need to be converted to a price per ounce. This conversion is done in the same manner as other conversions by multiplying the original cost by the appropriate measurement equivalent. In this case, since 16 ounces = 1 pound, $0.64 per pound is calculated to be equivalent to $0.04 per ounce.

$$\frac{\$0.64}{1\,\text{lb}} \times \frac{1\,\text{lb}}{16\,\text{oz}} = \frac{\$0.64 \times 1}{1 \times 16\,\text{oz}} = \frac{\$0.64}{16\,\text{oz}} = \frac{\textbf{\$0.04}}{\textbf{oz}}$$

The total cost of the sugar in the recipe can then be calculated by multiplying the EP amount (12 ounces) by the EP unit cost ($0.04 per ounce).

$$12 \text{ oz} \times \$0.04/\text{oz} = \textbf{\$0.48}$$

Checkpoint 6-2

Master Math™ Applications

1. Define unit cost.

_____

2. What is the AP unit cost of coffee per pound if a 5-pound bag of coffee costs $24.00?

_____

3. Explain the difference between an AP unit cost and an EP unit cost.

_____

_____

4. What is the EP unit cost of bell peppers if the AP unit cost is $1.10 per pound and the yield percent is 85%.

_____

5. If a recipe calls for 12 pounds of celery, what is the total cost of celery if the AP unit cost of celery is $0.80 per pound and the yield percentage is 70%?

_____

_____

CALCULATING AS-SERVED COSTS

Once the costs of the individual ingredients in a recipe are calculated, the costs are added to determine the total cost of the recipe. Then, as-served (AS) costs for that recipe can be calculated. An *as-served (AS) cost* is the cost of a menu item as it is served to a customer. It is important to note that an as-served (AS) cost is the total cost of the ingredients required to prepare *one* serving of a menu item. The exception would be a recipe with a yield equal to a single menu item, causing the cost of the recipe to be the same as the cost of the menu item.

Consider a chicken sandwich that is offered as a stand-alone item on a menu. If the sandwich consists of a grilled chicken breast served on a bun with a slice of tomato, a slice of cheese, lettuce, and mayonnaise, the costs of the ingredients are added to calculate the AS cost of the chicken sandwich. However, if the same chicken sandwich is offered on the menu as a meal that comes with fries and coleslaw, the AS cost of that menu item is equal to the total cost of the sandwich, fries, and coleslaw. On this menu, the chicken sandwich is treated as one ingredient of the chicken sandwich meal. **See Figure 6-3.**

Chicken Sandwich						
Ingredients	EP Quantity	EP Unit of Measure	AP Unit Cost	Yield Percentage	EP Unit Cost	Total Cost
Chicken Breast	4	oz	$0.20 per oz	90.0%	$0.22	$0.89
Bun	1	each	$0.40 per each	100.0%	$0.40	$0.40
Tomato Slice	1	oz	$0.18 per oz	90.0%	$0.20	$0.20
Cheese Slice	1	each	$0.20 per each	100.0%	$0.20	$0.20
Lettuce, shredded	0.25	oz	$0.10 per oz	80.0%	$0.13	$0.03
Mayonnaise	0.5	oz	$0.15 per oz	100.0%	$0.15	$0.08
					Total As-Served Cost:	$1.80

INDIVIDUAL RECIPE COST BECOMES PART OF MENU ITEM COST

Chicken Sandwich Meal						
Ingredients	EP Quantity	EP Unit of Measure	AP Unit Cost	Yield Percentage	EP Unit Cost	Total Cost
Chicken Sandwich	1	each	$1.80 per each	100.0%	$1.80	$1.80
French Fries	4	oz	$0.20 per oz	100.0%	$0.20	$0.80
Cole Slaw	3	oz	$0.20 per oz	100.0%	$0.20	$0.60
					Total As-Served Cost:	$3.20

Figure 6-3. As-served (AS) costs are the total costs of the ingredients and recipes that make up a menu item.

To calculate the AS cost of a menu item that is prepared in large quantities and then portioned into servings, divide the total cost of the recipe by the number of portions the recipe yields. For example, if the total cost of the ingredients to prepare 2 gallons of tomato soup is $15.00 and the recipe yields 32 8-fluid ounce servings, the AS cost of a serving of tomato soup would be $0.47.

$15.00 ÷ 32 servings = **$0.47/serving**

The AS cost of a beverage menu item is calculated in exactly the same way as the AS cost of a food menu item. For example, the AS cost of a chocolate milk shake could include the cost of vanilla ice cream, chocolate sauce, milk, whipped cream, and a cherry. Likewise, the AS cost of a cocktail could include the cost of soda or fruit juice, alcohol, and a garnish such as a slice of lemon or lime.

Wisconsin Milk Marketing Board

1. Define as-served (AS) cost.

_____

2. What is the AS cost of a taco made with $0.75 of ground beef, $0.24 of grated cheese, $0.08 of lettuce, $0.18 of tomatoes, and a $0.15 taco shell?

_____

3. If the taco in Question 2 is served as part of a meal that comes with black beans that costs $0.45 per serving and yellow rice that costs $0.35 per serving, what is the AS cost of the meal?

_____

4. What is the AS cost of a serving of tomato soup if the tomato soup recipe has a total ingredient cost of $50.00 and yields 25 servings?

_____

_____

CALCULATING FOOD COST PERCENTAGES

Another way of examining food costs is through the use of food cost percentages. A *food cost percentage* is a percentage that indicates how the cost of food relates to the menu prices and food sales of a foodservice operation. There are three types of food cost percentages used in the foodservice industry.

- **Menu-Item Food Cost Percentage.** A *menu-item food cost percentage* is the AS cost of a menu item divided by the menu price, written as a percent.

- **Overall Food Cost Percentage.** An *overall food cost percentage* is the total amount of money a foodservice operation spends on food divided by the total food sales over a defined period of time, written as a percent.

- **Target Food Cost Percentage.** A *target food cost percentage* is the percentage of food sales that a foodservice operation plans to spend on purchasing food.

Since beverage sales can be a large portion of a foodservice operation's total sales, beverage cost percentages are also used. Like a food cost percentage, a *beverage cost percentage* is a percentage that indicates how the cost of beverages relates to menu prices and beverage sales of a foodservice operation. The word beverage can be substituted for the word food in any of these calculations because the math is exactly the same.

Food cost percentages are only based on the costs of food and the sales of food. Likewise, beverage cost percentages are only based on the costs of beverages and the sales of beverages. If a menu item consists of both a food and beverage component that is going to be offered for a single price, the menu price must be based on a combined food and beverage cost percentage.

Menu-Item Food Cost Percentages

A menu-item food cost percentage is equal to the AS cost of a menu item divided by the menu price written as a percent. This percentage shows how the cost of the ingredients relates to the price charged for the menu item. To calculate a menu-item food cost percentage, apply the following formula:

IFC% = ASC ÷ MP

$$\text{Menu-Item Food Cost Percentage} = \frac{\text{AS Cost}}{\text{Menu Price}}$$

where

IFC% = menu-item food cost percentage

ASC = AS cost

MP = menu price

A foodservice operation earns money by charging more for the items on a menu than it costs to prepare those items. Therefore, the menu price is always higher than the AS cost of that item. Since a menu-item food cost percentage is equal to the AS cost divided by the menu price, a menu-item food cost percentage will always be less than 100%.

For example, if the menu price for the chicken sandwich shown in Figure 6-3 is $6.95 and the AS cost is $1.80, the menu-item food cost percentage for the chicken sandwich would be 25.9%.

IFC% = ASC ÷ MP

IFC% = $1.80 ÷ $6.95

IFC% = 0.259

IFC% = **25.9%**

Likewise, if the menu price for the chicken sandwich served with fries and coleslaw is $9.95 and the AS cost is $3.20, the menu-item food cost percentage for the meal would be 32.1%.

IFC% = ASC ÷ MP

IFC% = $3.20 ÷ $9.95

IFC% = 0.321

IFC% = **32.1%**

Idaho Potato Commission

Carlisle FoodService Products

Overall Food Cost Percentages

In addition to calculating food cost percentages for individual menu items, foodservice operations also calculate overall food cost percentages based on total food sales. An overall food cost percentage is equal to the total amount of money a foodservice operation spends on food divided by its total food sales over a defined period of time. To calculate an overall food cost percentage, use the following formula:

OFC% = FC ÷ FS

$$\text{Overall Food Cost Percentage} = \frac{\text{Total Food Costs}}{\text{Total Food Sales}}$$

where
OFC% = overall food cost percentage
FC = total food costs
FS = total food sales

For example, if a restaurant spends $9000 on food in a month when its total food sales were $36,000 for that month, the overall food cost percentage for the restaurant would be 25%.

OFC% = FC ÷ FS

OFC% = $9000 ÷ $36,000

OFC% = 0.25

OFC% = **25%**

The other 75% of the sales ($36,000 − $9000 = $27,000) would then be available for the foodservice operation to pay for expenses, such as payroll, rent, and utilities, and to contribute to the earnings of the owners of the operation. Lower food cost percentages translate into higher potential earnings for the owners of the operation.

Target Food Cost Percentages

It is common for a foodservice operation to establish a target food cost percentage for the operation. A target food cost percentage is the percentage of total food sales that a foodservice operation plans to spend on food. Foodservice owners and managers regularly compare overall food cost percentages with target food cost percentages to determine if food costs are higher or lower than planned.

When the overall food cost percentage of a foodservice operation is lower than the target food cost percentage, the operation has earned more money selling food than planned. Likewise, if the actual overall food cost percentage is higher than the target food cost percentage, the operation has earned less money selling food than planned. It is the responsibility of management to monitor food costs and make any necessary adjustments.

1. Define food cost percentage.

_____

_____

2. What is the menu-item food cost percentage of a salad with an AS cost of $1.75 and a menu price of $5.95? (Round answer to the tenths place.)

_____

_____

3. What is the overall food cost percentage of a restaurant that spends $400 a week on food when total food sales are $1800? (Round answer to the tenths place.)

_____

_____

4. Define target food cost percentage.

_____

_____

5. What is the cost percentage of a beverage item with an AS cost of $1.24 and a menu price of $6.00? (Round answer to the tenths place.)

_____

_____

CALCULATING MENU PRICES

There are three primary methods used by foodservice operations to calculate menu prices. These methods include food cost percentage pricing, perceived value pricing, and contribution margin pricing. Regardless of the pricing method used, the AS costs of the menu items must be known first.

Food Cost Percentage Pricing

Food cost percentage pricing begins by calculating a target price for each menu item. A *target price* is the price that a foodservice operation needs to charge for a menu item in order to meet its target food cost percentage. Target prices are calculated using the following formula:

$$TP = ASC \div TFC\%$$

where
TP = target price
ASC = AS cost of a menu item
$TFC\%$ = target food cost percentage

$$Target\ Price = \frac{AS\ Cost}{Target\ Food\ Cost\ Percentage}$$

For example, if the AS cost of a sandwich is $1.45 and the target food cost percentage is 30%, the target price would be $4.83.

$$TP = ASC \div TFC\%$$

$TP = \$1.45 \div 30\%$

$TP = \$1.45 \div 0.30$

$TP = \mathbf{\$4.83}$

Some foodservice operations will stop at this point and set the menu price equal to the target price. However, aside from some quick service restaurants, it is not common to see menu prices like $4.83. Most foodservice operations tend to consider target prices as a starting point and then adjust the price up or down based on how management thinks customers will perceive the price.

Perceived Value Pricing

Perceived value pricing is the process of adjusting a target menu price based on how management thinks a customer will perceive the price. Many foodservice operations use a combination of food cost percentage pricing and perceived value pricing to determine the final prices on a menu. **See Figure 6-4.**

There are two aspects to perceived value pricing. The first involves adjusting a target price to look like a price customers have become used to seeing on menus. For example, if a target price for a menu item is calculated to be $7.03, most foodservice operations would adjust the price down to $6.95 or $7.00. Likewise, if a target price for a menu item is calculated to be $7.74, most foodservice operations would adjust the price up to $7.95 or $8.00.

Therefore, the actual food cost percentage for the menu item would be slightly higher or lower than the target food cost percentage. For example, if the target price of a bowl of soup is calculated to be $4.67 based on an AS cost of $1.40 and a 30% target food cost, the restaurant might increase the price to $4.95. At $4.95, the actual food cost percentage would be 28.3%, which is slightly lower than the target food cost percentage of 30%.

$$IFC\% = ASC \div MP$$

$IFC\% = \$1.40 \div \$4.95$

$IFC\% = 0.283$

$IFC\% = \mathbf{28.3\%}$

The second aspect of perceived value pricing involves adjusting a target price so that customers will view the price as reasonable relative to other items on the menu and similar items sold elsewhere. For example, a restaurant may offer an appetizer of nachos made with tortillas chips, taco meat, cheese, and tomatoes that has an AS cost of $1.05. If the target price is calculated based on a 28% target food cost percentage, the target price would be $3.75.

TP = ASC ÷ TFC%

TP = $1.05 ÷ 28%

TP = $1.05 ÷ 0.28

TP = **$3.75**

Media Clips — Pricing Menu Items

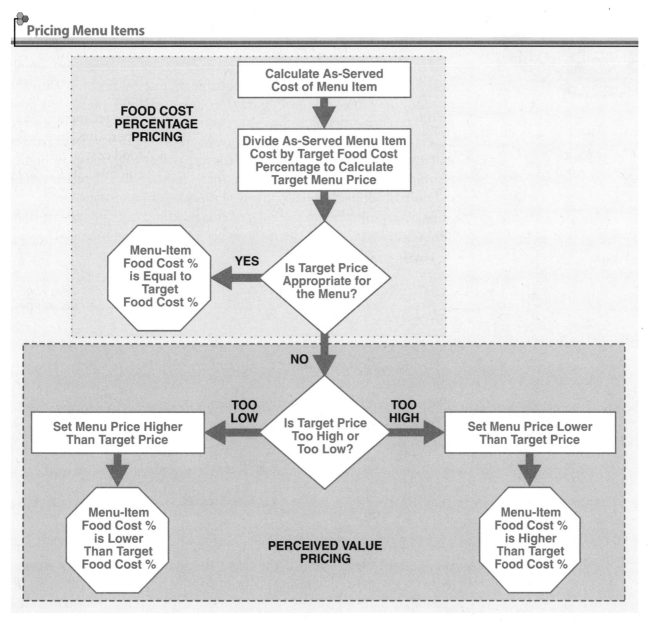

Figure 6-4. Many foodservice operations use a combination of food cost percentage pricing and perceived value pricing to calculate menu prices.

If other appetizers on the restaurant menu include buffalo wings for $8.95 and mozzarella sticks for $5.95, customers may be willing to pay more than $3.75 for the nachos because the perception is that nachos represent a substantial dish and should cost more than the mozzarella sticks but less than the buffalo wings. In a case like this, the restaurant is likely to increase the price of the nachos to $6.95, which means the menu-item food cost percentage for the nachos would be 15.1%.

IFC% = ASC ÷ MP

IFC% = $1.05 ÷ $6.95

IFC% = 0.151

IFC% = **15.1%**

If the price was left at the target price of $3.75 the restaurant would earn $2.70 ($3.75 – $1.05 = $2.70) on each order of nachos. With the price adjusted to $6.95 the restaurant would earn $5.90 ($6.95 – $1.05 = $5.90) on each order of nachos. A lower menu-item food cost percentage means that the foodservice operation earns more on each order of nachos that are sold. However, just because lower percentages result in higher earnings per order, it does not mean that a foodservice operation can or should base all menu prices on a lower food cost percentage.

To determine the price of other items based on the same menu-item food cost percentage as the nachos, use the following formula:

MP = ASC ÷ IFC%

where
MP = menu price
ASC = AS cost
IFC% = menu-item food cost percentage

$$\text{Menu Price} = \frac{\text{AS Cost}}{\text{Menu-Item Food Cost Percentage}}$$

For example, if the same restaurant selling the nachos prices a 24-ounce rib-eye steak with an AS cost of $14.00 based on the same 15.1% food cost percentage, the price of the steak would be $92.71.

MP = ASC ÷ IFC %

MP = $14.00 ÷ 15.1%

MP = $14.00 ÷ 0.151

MP = **$92.71**

It is unlikely that many customers would pay $92.71 for steak, even at a high-end restaurant. If the target food cost percentage is adjusted to 28%, the price of the steak would still be relatively high at $50.00 ($14.00 ÷ 0.28 = $50.00). Depending on the type of restaurant, customers may or may not perceive that price to be too high. If restaurant management thinks most customers will perceive a price of $50.00 as too high, the price may be reduced to be more in line with the perceived value of that restaurant's customers. If the final menu price set for the steak is $36.95, the food cost percentage will increase to 37.9%.

$$IFC\% = ASC \div MP$$

$IFC\% = \$14.00 \div \$36.95$

$IFC\% = 0.379$

$IFC\% = \mathbf{37.9\%}$

It is normal for items on a menu to have different food cost percentages. As long as there is a proper balance on a menu between items with lower food cost percentages and items with higher food cost percentages, the foodservice operation can maintain an overall food cost percentage that is close or equal to its target food cost percentage. **See Figure 6-5.**

Media Clips — Food Cost Percentages

Overall Food Cost Percentages

Daily Sales Report—Lunch						
Target Food-Cost Percentage = 30%						
Menu Item	AS Cost	Menu Price	Menu Item Food Cost % (AS Cost ÷ Menu Price)	Number Sold	Total Food Cost (AS Cost × Number Sold)	Total Food Sales (Menu Price × Number Sold)
Appetizers						
Bruschetta	$1.75	$5.95	29.4%	48	$84.00	$285.60
BBQ Shrimp	$4.25	$11.95	35.6%	27	$114.75	$322.65
Entrées						
Fried Chicken	$2.65	$12.95	20.5%	43	$113.95	$556.85
Strip Steak	$9.00	$25.95	34.7%	32	$288.00	$830.40
Desserts						
Strawberry Shortcake	$2.25	$5.95	37.8%	40	$90.00	$238.00
Chocolate Cake	$0.50	$3.95	12.7%	30	$15.00	$118.50
Total					**$705.70**	**$2,352.00**
Overall Total Food Cost Percentage (Total Food Costs ÷ Total Food Sales)						**30.0%**

Figure 6-5. If the food cost percentage for an individual menu item is higher or lower than the target food cost percentage, the overall food cost percentage can still be close or equal to the target food cost percentage.

Contribution Margin Pricing

Another method used to calculate menu prices is referred to as contribution margin pricing. A *contribution margin* is the amount added to the AS cost of a menu item to determine a menu price. A foodservice operation calculates prices with a contribution margin to ensure that the amount of money made on each serving of food will cover expenses and still result in a profit. This type of pricing is not common in restaurants but is often used when calculating a price charged per person for a special event.

For example, a hotel banquet operation is planning a party for 200 people. The menu includes a salad, an entrée, and a dessert that has a total AS cost of $20.00 per person. Therefore, the total AS cost for all of the food will be $4,000 (200 people × $20/person). The banquet manager calculates that all of the other expenses, such as labor and rental fees, total an additional $5000. It is hotel policy that a party of this size must generate earnings of $2000. Therefore, the contribution margin for this event would be $7000 ($5000 + $2000).

To calculate the contribution margin on a per person basis, the banquet manager simply divides the total contribution margin by the number of people attending the event.

$7000 ÷ 200 people = **$35/person**

This contribution margin is then added to the AS cost per person to calculate the final menu price for the party, which is $55 per person.

$20.00/person + $35.00/person = **$55/person**

Beverage Pricing

Beverage pricing is calculated the same way as food pricing. Simply substitute beverage costs for food costs in the original calculation for menu-item cost percentages as follows:

IBC % = ASC ÷ MP

where
IBC% = menu-item beverage cost percentage
ASC = AS cost
MP = menu price

$$\text{Menu Item Beverage Cost Percentage} = \frac{\text{AS Cost}}{\text{Menu Price}}$$

For example, the beverage cost percentage of a cocktail with an AS cost of $1.65 that is priced at $7.95 has a beverage cost percentage of 20.8%.

IBC% = ASC ÷ MP

IBC% = $1.65 ÷ $7.95

IBC% = 0.208

IBC% = **20.8%**

If a restaurant has established a target beverage cost percentage of 25%, a bottle of iced tea with an AS cost of $1.15 would have a price of $4.60. The target price of the iced tea is calculated using the following formula:

TP = ASC ÷ TBC%

where
TP = target price
ASC = AS cost
TBC% = target beverage cost percentage

$$\text{Target Price} = \frac{\text{AS Cost}}{\text{Target Beverage Cost Percentage}}$$

TP = ASC ÷ TBC%

$TP = \$1.15 ÷ 25\%$

$TP = \$1.15 ÷ 0.25$

$TP = \textbf{\$4.60}$

Charlie Trotter's

Just as with food, the price for the bottle of beer will likely be increased to $4.95 or $5.00 depending on the pricing philosophy of the restaurant. The price may be adjusted even further based on the perceived value of the beer relative to other beverage items offered on the menu.

Finally, if a foodservice operation had beverage costs of $2500 and beverage sales of $10,000 in a given month, the overall beverage cost percentage could be calculated using the following formula:

OBC% = BC ÷ BS

where
OBC% = overall beverage
          cost percentage
BC = total beverage costs
BS = total beverage sales

$$\text{Overall Beverage Cost Percentage} = \frac{\text{Total Beverage Costs}}{\text{Total Beverage Sales}}$$

OBC% = BC ÷ BS

$OBC\% = \$2500 ÷ \$10,000$

$OBC\% = 0.25$

$OBC\% = \textbf{25\%}$

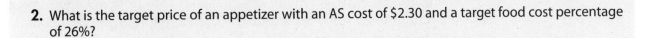

Checkpoint 6-5

Master Math™
Applications

1. Define target price.

_____

2. What is the target price of an appetizer with an AS cost of $2.30 and a target food cost percentage of 26%?

_____

_____

3. How much money is earned from selling a dessert with an AS cost of $2.15 and a food cost percentage of 32%?

_____

_____

4. Define perceived value pricing.

_____

_____

5. Explain what happens to the food cost percentage of a menu item when the price of the menu item is increased?

_____

6. Define contribution margin.

_____

_____

7. If the total contribution margin for a catered event for 50 people is calculated to be $500 and the AS cost of the food for the event is $10.00 per person, how much should the catering department charge per person for the event?

_____

_____

8. What is the target price for bottled water that has an AP cost of $0.34 per bottle if the target beverage cost percentage is 22%?

_____

_____

9. What is the overall beverage cost percentage for a foodservice operation that had beverage costs of $275 in a week when the total beverage sales were $950?

_____

_____

USING PRICING FORMS

A *pricing form* is a tool often used to help calculate the AS cost of a menu item and establish a menu price. They may vary from one foodservice operation to another. **See Figure 6-6.** Pricing forms generally contain the following elements:

- **Menu Item Name**—The name on the pricing form should match the name used on the menu.
- **Ingredients**—The list of ingredients required to prepare the menu item.
- **Edible-Portion (EP) Quantity**—The EP quantity is the amount of each ingredient required to prepare the menu item.
- **Number of Portions**—Lists the number of portions that are made with the ingredients listed on the pricing form.

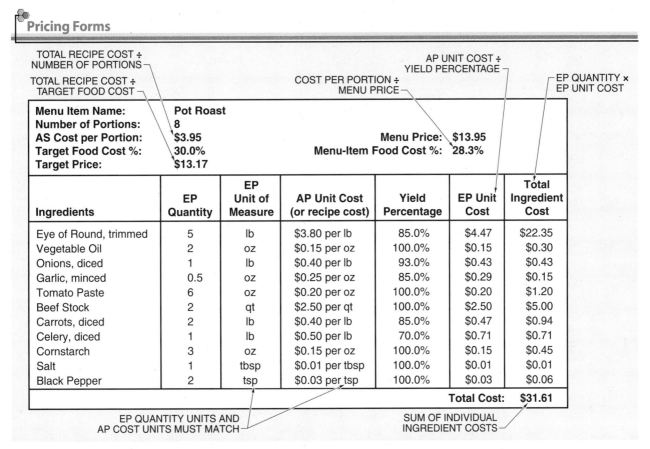

TOTAL RECIPE COST ÷
NUMBER OF PORTIONS

TOTAL RECIPE COST ÷
TARGET FOOD COST

AP UNIT COST ÷
YIELD PERCENTAGE

COST PER PORTION ÷
MENU PRICE

EP QUANTITY ×
EP UNIT COST

Menu Item Name:	Pot Roast					
Number of Portions:	8					
AS Cost per Portion:	$3.95					
Target Food Cost %:	30.0%		Menu Price:	$13.95		
Target Price:	$13.17		Menu-Item Food Cost %:	28.3%		

Ingredients	EP Quantity	EP Unit of Measure	AP Unit Cost (or recipe cost)	Yield Percentage	EP Unit Cost	Total Ingredient Cost
Eye of Round, trimmed	5	lb	$3.80 per lb	85.0%	$4.47	$22.35
Vegetable Oil	2	oz	$0.15 per oz	100.0%	$0.15	$0.30
Onions, diced	1	lb	$0.40 per lb	93.0%	$0.43	$0.43
Garlic, minced	0.5	oz	$0.25 per oz	85.0%	$0.29	$0.15
Tomato Paste	6	oz	$0.20 per oz	100.0%	$0.20	$1.20
Beef Stock	2	qt	$2.50 per qt	100.0%	$2.50	$5.00
Carrots, diced	2	lb	$0.40 per lb	85.0%	$0.47	$0.94
Celery, diced	1	lb	$0.50 per lb	70.0%	$0.71	$0.71
Cornstarch	3	oz	$0.15 per oz	100.0%	$0.15	$0.45
Salt	1	tbsp	$0.01 per tbsp	100.0%	$0.01	$0.01
Black Pepper	2	tsp	$0.03 per tsp	100.0%	$0.03	$0.06
					Total Cost:	$31.61

EP QUANTITY UNITS AND
AP COST UNITS MUST MATCH

SUM OF INDIVIDUAL
INGREDIENT COSTS

Figure 6-6. A pricing form is used to help calculate the AS cost of a menu item and establish a menu price.

- **As-Purchased (AP) Unit Cost**—The unit cost for each ingredient in the same unit of measure as the EP quantity of each ingredient.
- **Yield Percentage**—The yield percentage of each ingredient listed.
- **Edible-Portion (EP) Unit Cost**—Calculated by dividing the AP unit cost of each ingredient by the ingredient's yield percentage.
- **Total Ingredient Cost**—The total cost for each ingredient is calculated by multiplying the EP quantity of each ingredient by the EP unit cost of each ingredient.
- **Total Cost**—The sum of all of the individual ingredient costs.
- **AS Cost per Portion**—Calculated by dividing the total cost by the number of portions.
- **Target Food Cost Percentage**—The target food cost percentage established by the foodservice operation.
- **Target Price**—Calculated by dividing the AS cost per portion by the target food cost percentage.
- **Menu Price**—The price listed on the menu for a single portion of the menu item.
- **Menu-Item Food Cost Percentage**—Calculated by dividing the AS cost per portion by the menu price.

Forms and Tables

The procedure for using a pricing form is as follows:

1. Enter information from the standardized recipe including the menu item name, number of portions, ingredients, and ingredient amounts (EP quantities). **See Figure 6-7.**

Menu Item Name:	Pot Roast					
Number of Portions:	8					
AS Cost per Portion:						
Target Food Cost %:						
Target Price:						

Menu Price:
Menu-Item Food Cost %:

Ingredients	EP Quantity	EP Unit of Measure	AP Unit Cost (or recipe cost)	Yield Percentage	EP Unit Cost	Total Ingredient Cost
Eye of Round, trimmed	5	lb				
Vegetable Oil	2	oz				
Onions, diced	1	lb				
Garlic, minced	0.5	oz				
Tomato Paste	6	oz				
Beef Stock	2	qt				
Carrots, diced	2	lb				
Celery, diced	1	lb				
Cornstarch	3	oz				
Salt	1	tbsp				
Black Pepper	2	tsp				
					Total Cost:	

Figure 6-7. The first step in completing a pricing form is to enter the information from the recipe being priced.

2. Record the AP unit cost of each ingredient. If necessary, convert the unit cost to the same unit of measure as used in the EP quantity.
3. Record the yield percentage for each ingredient.
4. Calculate the EP unit costs by dividing the AP unit cost of each ingredient by the yield percentage of the ingredient.
5. Calculate the total cost of each ingredient by multiplying the EP quantity of each ingredient by the EP unit cost of each ingredient. **See Figure 6-8.**
6. Calculate the total cost by adding the costs of each ingredient.
7. Calculate the cost per portion by dividing the total recipe cost by the number of portions.
8. Calculate the target price by dividing the total cost by the target food cost percentage.
9. Adjust the target price if desired and record the menu price.
10. Calculate the food cost percentage by dividing the cost per portion by the menu price. **See Figure 6-9.**

Menu Item Name:	Pot Roast					
Number of Portions:	8					
AS Cost per Portion:				Menu Price:		
Target Food Cost %:				Menu-Item Food Cost %:		
Target Price:						

Ingredients	EP Quantity	EP Unit of Measure	AP Unit Cost (or recipe cost)	Yield Percentage	EP Unit Cost	Total Ingredient Cost
Eye of Round, trimmed	5	lb	$3.80 per lb	85.0%	$4.47	$22.35
Vegetable Oil	2	oz	$0.15 per oz	100.0%	$0.15	$0.30
Onions, diced	1	lb	$0.40 per lb	93.0%	$0.43	$0.43
Garlic, minced	0.5	oz	$0.25 per oz	85.0%	$0.29	$0.15
Tomato Paste	6	oz	$0.20 per oz	100.0%	$0.20	$1.20
Beef Stock	2	qt	$2.50 per qt	100.0%	$2.50	$5.00
Carrots, diced	2	lb	$0.40 per lb	85.0%	$0.47	$0.94
Celery, diced	1	lb	$0.50 per lb	70.0%	$0.71	$0.71
Cornstarch	3	oz	$0.15 per oz	100.0%	$0.15	$0.45
Salt	1	tbsp	$0.01 per tbsp	100.0%	$0.01	$0.01
Black Pepper	2	tsp	$0.03 per tsp	100.0%	$0.03	$0.06
					Total Cost:	

Figure 6-8. The cost of each ingredient in a menu item is calculated by multiplying the EP quantity by the EP unit cost of each ingredient.

Menu Item Name:	Pot Roast					
Number of Portions:	8					
AS Cost per Portion:	$3.95			Menu Price:	$13.95	
Target Food Cost %:	30.0%			Menu-Item Food Cost %:	28.3%	
Target Price:	$13.17					

Ingredients	EP Quantity	EP Unit of Measure	AP Unit Cost (or recipe cost)	Yield Percentage	EP Unit Cost	Total Ingredient Cost
Eye of Round, trimmed	5	lb	$3.80 per lb	85.0%	$4.47	$22.35
Vegetable Oil	2	oz	$0.15 per oz	100.0%	$0.15	$0.30
Onions, diced	1	lb	$0.40 per lb	93.0%	$0.43	$0.43
Garlic, minced	0.5	oz	$0.25 per oz	85.0%	$0.29	$0.15
Tomato Paste	6	oz	$0.20 per oz	100.0%	$0.20	$1.20
Beef Stock	2	qt	$2.50 per qt	100.0%	$2.50	$5.00
Carrots, diced	2	lb	$0.40 per lb	85.0%	$0.47	$0.94
Celery, diced	1	lb	$0.50 per lb	70.0%	$0.71	$0.71
Cornstarch	3	oz	$0.15 per oz	100.0%	$0.15	$0.45
Salt	1	tbsp	$0.01 per tbsp	100.0%	$0.01	$0.01
Black Pepper	2	tsp	$0.03 per tsp	100.0%	$0.03	$0.06
					Total Cost:	$31.61

Figure 6-9. A target price is determined before setting a menu price and calculating the menu-item food cost percentage.

After a pricing form has been completed for all of the items on a menu, the foodservice operation updates the pricing forms when the cost of ingredients change to ensure that menu prices are adjusted as needed.

Checkpoint 6-6

1. Define pricing form.

_____

_____

2. What are the two values that must be known in order to calculate the total ingredient cost of each ingredient in a menu item?

_____

3. Explain the difference between the target price and the menu price on a pricing form.

_____

_____

Quick Quiz® Chapter 6

Flash Cards

Chapter 6 Summary

The most fundamental food and beverage cost is the as-purchased (AP) cost. AP costs are then broken down into unit costs to calculate the costs of ingredients in a recipe. For food products that are trimmed before being used in a recipe, AP unit costs are converted to edible portion (EP) unit costs based on the yield percentage of the product. The total cost of the ingredients required to prepare a menu item is referred to as the as-served (AS) cost.

After the AS cost of a menu item is calculated, menu prices can be established. A menu-item cost percentage indicates how the cost of preparing a menu item relates to the item's menu price. Most foodservice operations use target food and beverage cost percentages as tools to help calculate menu prices. Calculated menu prices are often adjusted based on perceived customer value.

Some foodservice operations calculate menu prices by adding contribution margins to the costs of food and beverages. Overall cost percentages are regularly compared to target cost percentages. The comparison identifies increased food costs to determine if menu prices need to be adjusted. Food costs and menu pricing information are documented on pricing forms and updated as AP costs change.

Cres Cor

Checkpoint Answers

Checkpoint 6-1

1. An as-purchased (AP) cost is the original amount paid for a product in the form it was ordered and received.
2. An AP cost can be found on the invoice from a supplier.
3. $15
4. $18

Checkpoint 6-2

1. A unit cost is the cost of a product per unit of measure.
2. $4.80/lb   ($24.00 ÷ 5 lb = $4.80/lb)
3. The AP unit cost of a product is based on the form in which it is ordered and received. The EP unit cost of a product reflects the cost of the product after determining the cost of waste generated by trimming the product.
4. $1.29/lb   ($1.10 ÷ 0.85 = $1.29)
5. $13.68   ($0.80/lb ÷ 0.7 = $1.14/lb, and $1.14/lb × 12 lb = $13.68)

Checkpoint 6-3

1. An as-served (AS) cost is the cost of a menu item as it is served to a customer.
2. $1.40   ($0.75 + $0.24 + $0.08 + $0.18 + $0.15 = $1.40)
3. $2.20   ($1.40 + $0.45 + $0.35 = $2.20)
4. $2.00/serving   ($50.00 ÷ 25 servings = $2.00/serving)

Checkpoint 6-4

1. A food cost percentage is a mathematical indication of how the cost of purchasing food relates to the prices of menu items and the food sales of a foodservice operation.
2. 29.4%   ($1.75 ÷ $5.95 = 0.294 = 29.4%)
3. 22.2 %   ($400 ÷ $1800 = 0.222 = 22.2%)
4. A target food cost percentage is the percentage of food sales that a foodservice operation plans to spend on food.
5. 20.7%   ($1.24 ÷ $6.00 = 0.207 = 20.7%)

continued . . .

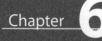
Checkpoint 6-5

1. A target price is the price calculated by dividing the AS cost of a menu item by a target food-cost percentage.

2. $8.85   ($2.30 ÷ 0.26 = $8.85)

3. $4.57   ($2.15 ÷ 0.32 = $6.72, and $6.72 - $2.15 = $4.57)

4. Perceived value pricing is the process of adjusting a target menu price based on how management thinks a customer will perceive the price of the menu item.

5. The food cost percentage of a menu item decreases when the price of the menu item increases.

6. A contribution margin is the amount added to an AS cost of a menu item in order to determine a menu price.

7. $20.00/person   ($500 ÷ 50 people = $10.00/person, and $10.00/person + $10.00/person = $20.00/person)

8. $1.55   ($0.34 ÷ 0.22 = $1.55)

9. 28.9%   ($275 ÷ $950 = 0.289 = 28.9%)

Checkpoint 6-6

1. A pricing form is a tool used to help calculate the AS cost of a menu item and establish a menu price.

2. To calculate the total cost of each ingredient in a menu item, the EP quantity and the EP cost of the ingredient must be known.

3. A target price is the price calculated based on a target food cost percentage. The menu price is the final price provided on the menu.

Best western
160.00

Radisson comfort inn
14.99
180

holiday inn
174.95

Calculating Revenue and Expenses

In order to determine whether a foodservice operation is financially successful, the amount of money coming into the operation and the amount of money leaving the operation must be known. Foodservice workers need to have a working knowledge of the tools and processes used for documenting and calculating the sale of food and beverages to customers. Likewise, foodservice workers need to understand the different types of expenses that are incurred by a foodservice operation and how those expenses relate to the financial success of the operation.

Chapter Objectives

1. Explain how a guest check is processed.
2. Calculate discounts, sales taxes, guest check totals, and gratuities.
3. Explain the purpose of a point-of-sale (POS) device.
4. Demonstrate how to return change to a customer who pays with cash.
5. Explain how daily sales revenue is calculated and recorded.
6. Describe the different expense categories of a foodservice operation.
7. Calculate the value of inventory.
8. Calculate the cost of goods sold.
9. Explain how payroll expenses are calculated.
10. Explain the difference between variable and fixed expenses.

Key Terms

- revenue
- discount
- sales tax
- point-of-sale (POS) device
- gratuity
- capital expense
- cost of goods sold
- inventory
- operating expense
- payroll expense
- gross pay
- net pay
- interest
- variable expense
- fixed expense

CALCULATING REVENUE

Determining whether a foodservice operation is financially successful starts with calculating revenue. *Revenue* is the total amount of money received by a foodservice operation from sales to customers. The process of calculating revenue begins when a customer places an order. A foodservice operation must be able to document what has been ordered by the customer, provide the customer with a total cost for that order, and properly collect and record payments.

Cres Cor

Processing Guest Checks

One of the most common tools that a foodservice operation uses to document what is sold to a customer is a guest check. A guest check lists the items ordered by a customer, the prices to be charged for those items, and the total amount of money due. Guest checks can be written by hand or generated and printed by a computer or a cash register.

In addition to the amount due for food and beverages, the guest check may also include additional charges for items such as sales tax and service charges. It may also reflect discounts that reduce the total amount owed based on coupons or other promotions offered by the foodservice operation. The processing of a guest check ends once payment has been received by the foodservice operation and the proper amount of change and a receipt for payment have been returned to the customer.

Writing Guest Checks. Handwritten guest checks are most commonly used in restaurants where customers place an order from a table or at a counter. Servers frequently use handwritten guest checks to record customer orders. **See Figure 7-1.** Often, the guest check is a form that generates one copy for the server and one for the kitchen. The designs of handwritten guest checks vary but most contain the following common information:

- **General Information.** General information includes items such as the date, table number, server's identification number or name, and the number of guests (customers) at the table. Also, most guest checks include preprinted numbers so that each guest check is unique and easy to track.

- **Main Body.** The main body of the guest check is used to record the menu items ordered and the quantity and cost of each item.

- **Subtotal.** The subtotal of a guest check is the sum of all food and beverage charges. A subtotal may be increased or decreased before arriving at a grand total.

- **Discounts.** A *discount* is a percentage or fixed amount by which the subtotal is reduced due to a promotion offered by the foodservice operation. For example, a restaurant may offer 10% off all appetizers on Tuesdays or print coupons that offer $3.00 off any large pizza.

- **Sales Taxes.** A *sales tax* is a fee that must be collected from customers based on the requirements of state and local governments where the foodservice operation is located. These fees are then paid to the government-taxing agency at a later date. In most cases, sales taxes are calculated based on a percentage of the amount charged for food and beverages minus any discounts.

- **Total.** The total on a guest check is the final amount of money due from the customer. The total is calculated by subtracting any discounts from the subtotal and then adding the sales tax.

Guest Check

Media Clips — Processing Guest Checks

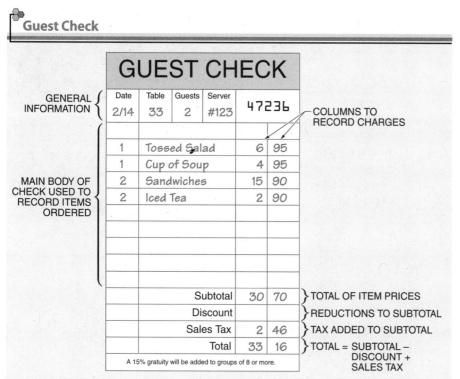

Figure 7-1. Handwritten guest checks are one of the most common tools used to document sales in a foodservice operation.

In order to process a handwritten guest check, certain skills are required such as addition, subtraction, and multiplication. The first step in processing a guest check involves filling in the general information. Next, the items ordered are recorded in the main body.

For example, two customers come into a pizza restaurant for lunch and are seated at table 20. Before approaching the table, the server will write the date, table number (20), number of guests (2), and his or her name or identification number on the guest check. The server writes down additional information as the customers place their orders. The first customer orders an individual pizza with pepperoni, a house salad, and a soft drink. The second customer orders an individual pizza with sausage, green peppers, and onions; a cup of minestrone soup; and a soft drink.

After recording the order, the server may give a copy of the guest check to workers in the kitchen. Sometimes the original guest check is passed off into the kitchen and then returned to the server once the food is prepared. **See Figure 7-2.**

Recording and Placing an Order

GUEST CHECK				
Date	Table	Guests	Server	54896
9/15	20	2	#123	
1	Individual Pizza			
	pepperoni			
1	House Salad			
1	Individual Pizza			
	sausage, green			
	peppers, onions			
1	Minestrone Soup			
2	Soft Drinks			
	Subtotal			
	Discount			
	Sales Tax			
	Total			
A 15% gratuity will be added to groups of 8 or more.				

Carlisle FoodService Products

Figure 7-2. A copy of the guest check is sometimes created for use in the kitchen.

Before presenting the final check to the customer for payment, the server records the cost of each menu item ordered. **See Figure 7-3.** If an entry on the guest check is a single menu item, such as one house salad, the menu price is simply written in the column where the charges are recorded. If an entry includes more than one of the same menu item, such as two soft drinks, the menu price must be multiplied by the number of items ordered and then the product is recorded in the charges column. Next, all of the charges are added to calculate the subtotal of charges for food and beverages.

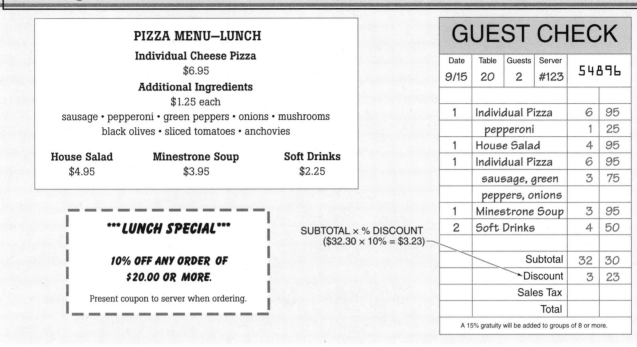

Figure 7-3. The subtotal on a guest check is the sum of the charges for all food and beverage items on an order and is sometimes reduced by a discount.

Calculating Discounts. If the foodservice operation is offering a promotion, such as discount coupons, the amount of the discount needs to be determined and subtracted from the subtotal. For example, if a pizza restaurant offers a coupon for 10% off any order of $20.00 or more, the server would multiply the subtotal by 10% and then record the product on the guest check.

Calculating Sales Tax. The next step is to calculate the sales tax. Sales taxes are based on a percentage of the actual charges to the customer for food and beverages. Therefore, any discount must be subtracted from the subtotal before the sales tax can be calculated. In the pizza example, the subtotal is $32.30 and the discount is $3.23, so the actual amount being charged for food and beverages is $29.07 ($32.30 − $3.23 = $29.07). If the sales tax where the pizza restaurant is located is 8%, the tax is calculated by multiplying $29.07 by 8%.

Sales Tax = $29.07 × 8%

Sales Tax = $29.07 × 0.08

Sales Tax = $2.325 = **$2.33**

The total on the guest check is calculated by subtracting the discount from the subtotal and then adding the sales tax. **See Figure 7-4.**

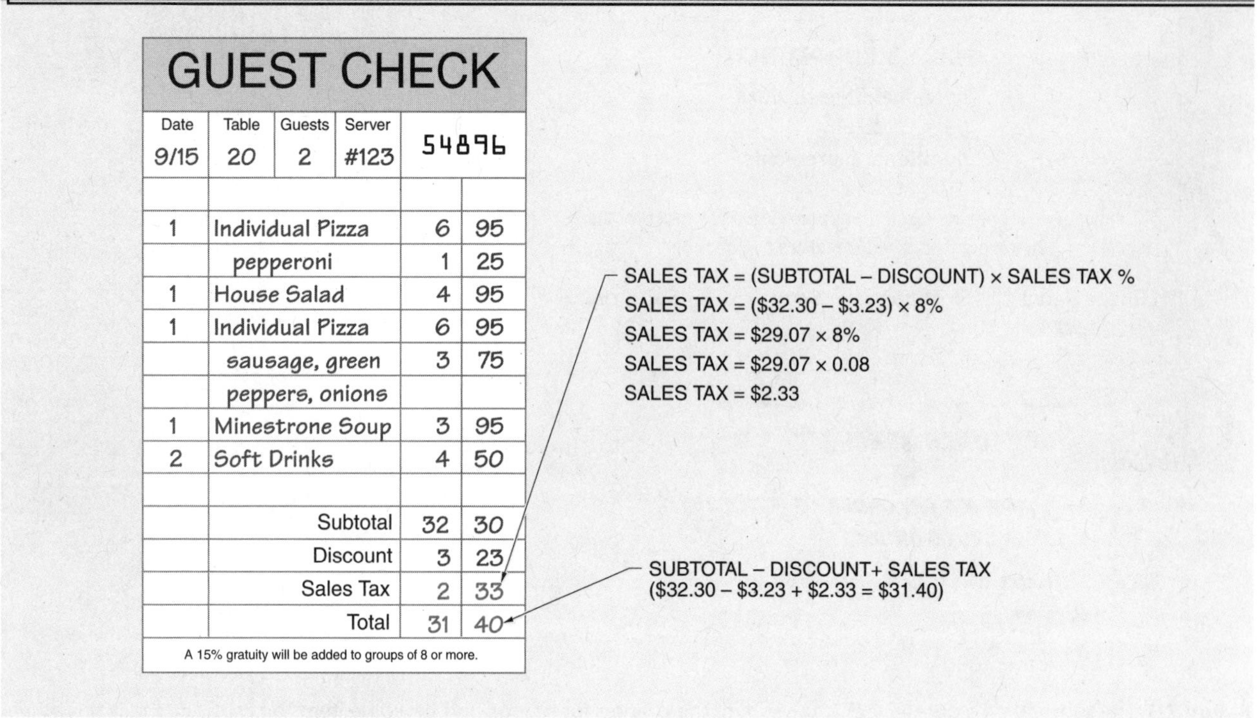

GUEST CHECK

Date	Table	Guests	Server	54896	
9/15	20	2	#123		
1	Individual Pizza			6	95
	pepperoni			1	25
1	House Salad			4	95
1	Individual Pizza			6	95
	sausage, green			3	75
	peppers, onions				
1	Minestrone Soup			3	95
2	Soft Drinks			4	50
	Subtotal			32	30
	Discount			3	23
	Sales Tax			2	33
	Total			31	40

A 15% gratuity will be added to groups of 8 or more.

SALES TAX = (SUBTOTAL − DISCOUNT) × SALES TAX %
SALES TAX = ($32.30 − $3.23) × 8%
SALES TAX = $29.07 × 8%
SALES TAX = $29.07 × 0.08
SALES TAX = $2.33

SUBTOTAL − DISCOUNT+ SALES TAX
($32.30 − $3.23 + $2.33 = $31.40)

Figure 7-4. The total amount of a guest check is equal to the food and beverage subtotal minus discounts plus sales tax.

Using Point-of-Sale Devices. Most foodservice operations use electronic point-of-sale devices or systems to process guest checks. A *point-of-sale (POS) device* is an electronic tool used to process customer orders, print guest checks, track customer and financial information, and generate financial reports. A POS device can be as simple as a single cash register or as sophisticated as several POS devices located throughout a restaurant that are linked to one another (referred to as a POS system). **See Figure 7-5.**

Sometimes handwritten guest checks are used in conjunction with a cash register or POS device. A POS device is used to keep track of the amount of money received from customers and to help calculate the change that is returned to a customer who pays with cash. Some cash registers can be programmed with menu item names and prices and the server may enter customer orders into the cash register. The register then prints guest checks electronically. **See Figure 7-6.**

The most sophisticated POS systems not only allow the server to enter customer orders and generate computer-generated guest checks, but also have the ability to send the orders to a printer or video display located in the kitchen. This eliminates the need for the server to provide the kitchen with a hard copy of each guest check.

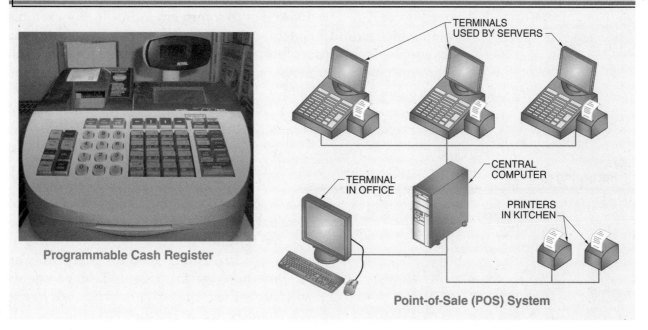

TERMINALS USED BY SERVERS

CENTRAL COMPUTER

TERMINAL IN OFFICE

PRINTERS IN KITCHEN

Programmable Cash Register

Point-of-Sale (POS) System

Figure 7-5. POS equipment can be as simple as a programmable cash register or as sophisticated as a group of devices linked to a central computer.

These POS systems usually have the ability to generate reports that management can use to see how much of each menu item is sold over a period of time. This information may help management make decisions regarding staffing needs and ordering requirements. The information is also useful for making judgments regarding which menu items sell better than others or sell better during certain times of the year.

Even though POS systems help eliminate errors by automatically performing the math calculations required to generate a guest check, the server must still know how to spot a mistake. The information generated by a POS system is only as good as the information entered into the system. For example, if the server accidentally enters the wrong item or the wrong number of menu items and does not realize it, the mistake can cause problems in the kitchen as well as errors on the guest check.

Collecting Payment and Returning Change. When a customer pays with cash, it is often necessary to return change. The amount of change given to the customer should be equal to the difference between the amount of cash presented for payment and the total amount of the guest check. For example, if a customer uses a $20 bill to pay for a guest check with a total of $13.75, the customer is returned $6.25 in change ($20.00 − $13.75 = $6.25).

Electronically Generated Guest Check

5/12/2012 9:06:57 PM		Receipt#: 18567	
	Mom's Bistro		
DESCRIPTION	QTY	PRICE	EXT PRC
# 301	1	$6.00	$6.00
Crab Cakes Appetizer			
# 303	1	$5.00	$5.00
Onion Tartlet Appetizer			
# 310	1	$4.00	$4.00
Potato Leek w/ Wild Rice			
# 311	1	$4.00	$4.00
Italian Chicken Noodle			
# 322	1	$14.00	$14.00
Glazed Pork Tender			
# 328	1	$10.00	$10.00
Chix Pot Pie			
# 10000	4	$0.85	$3.40
Soft Drink - 12 oz. Can			
# 371	2	$3.00	$6.00
Red Velvet Cup Cake			
8 Item(s)	Subtotal:		$52.40
	10.00 % Disc:		- $5.24
	7.75 % Tax:		$3.65
	RECEIPT TOTAL:		**$50.81**
		Total Savings $ $5.24	

Thank you for visiting Mom's Bistro.

Figure 7-6. A POS device, or system, can generate and print guest checks.

Instead of simply handing the customer the change all at once, it is appropriate to return it to the customer in an orderly process. **See Figure 7-7.** The first step is to return any coins. In this example the cashier would hand the customer a quarter ($0.25) and say "twenty-five cents makes fourteen dollars." The next step would be to hand the customer a $1.00 bill and say "fifteen," and then hand the customer a $5.00 bill and say "and five makes twenty." The process always ends with the total amount of money originally presented by the customer for payment. This process helps to ensure that the proper amount of change is returned to the customer.

In addition to cash, most foodservice operations offer customers the option of paying with a credit or debit card. Some restaurants even have their own gift cards that can be used for payment. The method for processing card payments depends on the type of system used by the foodservice operation. Sometimes cards are processed through a separate, stand-alone terminal while others are processed through a POS system.

Calculating Gratuities. In a foodservice operation where a server waits on the customer, it is customary to leave a gratuity, or tip. A *gratuity* is the amount of money left by a customer as thanks for the services rendered. The amount of a gratuity is usually left to the discretion of the customer. However, many restaurants and most banquet facilities include a gratuity as part of the check for larger groups (typically more than six or eight people). A restaurant's gratuity policy is usually provided on the menu. A banquet facility will include the gratuity in the customer contract.

Although there are no minimum or maximum requirements when calculating gratuities, it has become widely accepted that a gratuity in a full-service restaurant should be between 15% and 20% of the food and beverage total on the guest check. It is not necessary to consider any discounts or sales tax when calculating a gratuity.

Returning Change

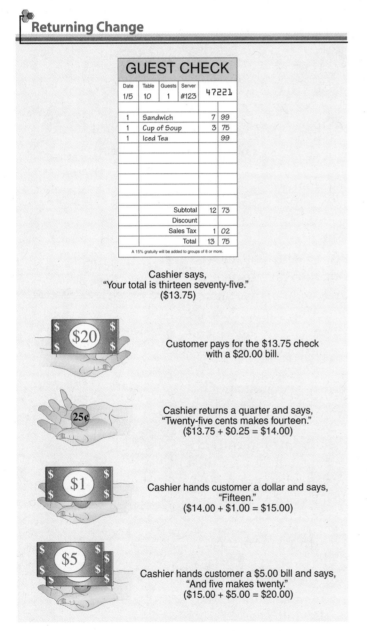

Figure 7-7. Returning change to a customer in an orderly fashion helps to minimize the chance of returning the incorrect amount of change.

For example, a 15% gratuity on a check with a food and beverage subtotal of $45.00 would be $6.75.

$$\$45.00 \times 15\% = \$45.00 \times 0.15 = \mathbf{\$6.75}$$

A 20% gratuity on the same check would be $9.00.

$$\$45.00 \times 20\% = \$45.00 \times 0.20 = \mathbf{\$9.00}$$

Based on this example, if the customer leaves a gratuity between $6.75 and $9.00 it indicates that the customer was satisfied with the level of service received. Likewise, if the gratuity left is significantly more than $9.00, it indicates that the customer was extremely satisfied with the level of service.

Gratuity Guide

To calculate a 15% gratuity, multiply the food and beverage subtotal by 10%. Then, divide that result by 2 to calculate 5% of the total. Finally, add the two amounts to arrive at 15%.

For example, a 15% gratuity on a $50.00 check would be calculated as follows:

10% of $50.00 = $50.00 = $5.00

5% of $50.00 = $5.00 ÷ 2 = $2.50

15% of $50.00 = $5.00 + $2.50 = $7.50

Calculating Sales Revenue

Most foodservice operations calculate sales revenue on a daily basis. The calculations are documented using a standard form, such as a daily sales record. **See Figure 7-8.** The name of the form and the specific information recorded will vary based on the needs of the particular operation, but will always contain at least the following basic information:

- **Total Sales.** Total sales is the sum of all of the individual guest check food and beverage subtotals minus all discounts. Foodservice operations that utilize programmable cash registers or POS systems can easily generate reports that break the total sales down into further categories such as sales of food and sales of beverages.

- **Gross Sales.** Gross sales is the sum of the totals (after discounts) on all of the guest checks processed that day, which is equal to the total sales plus the total sales tax.

- **Cash at End of Day.** The total amount of cash at the end of the day is determined by physically counting all of the cash in the cash register.

- **Cash at Start of Day.** The amount of cash available at the start of the day is primarily used for returning change to customers paying with cash.

- **Total Cash Revenue.** Total cash revenue is the difference between the "cash at end of day" and the "cash at start of day." This is the total amount of cash received from customers that day.

- **Total Card Revenue.** Total card revenue is the total amount of credit, debit, or gift card charges received from customers that day.

- **Total Revenue.** Total revenue is the total cash revenue added to the total card revenue. This total should also match the gross sales.

- **Over (Short).** If the total revenue is more than the gross sales, then too much money was received from customers, referred to as "over". If the total revenue is less than the gross sales, then too little money was received from customers, referred to as "short." The difference between the total revenue and the gross sales should be zero.

Daily Sales Record	
Total Sales	$ 14,536.89
Total Food Sales	$ 11,457,60
Total Beverage Sales	$ 3079.29
Sales Tax (8%)	$ 1162.95
Gross Sales (*Total Sales + Sales Tax*)	$ 15,699.40
Cash at End of Day	$ 5638.25
Cash at Start of Day	$ 400.00
Total Cash Revenue (*Cash at End of Day – Cash at Start of Day*)	$ 5238.25
Total Card Revenue	$ 10,461.15
Credit Cards	$ 2653.12
Debit Cards	$ 5893.35
Gift Cards	$ 1914.68
Total Revenue (*Total Cash Revenue + Total Card Revenue*)	$ 15,699.40
Over (Short) (*Total Revenue – Gross Sales*)	$ 0.00
Report completed by: _____ Signature/Date	

INFORMATION FROM GUEST CHECKS, CASH REGISTER, OR POS SYSTEM REPORTS

GROSS SALES SHOULD EQUAL TOTAL REVENUE

Figure 7-8. The daily gross sales of an operation (amount charged to customers) should equal the total revenue (amount of money received from customers).

Forms and Tables

In addition to calculating and documenting the total revenue for the day, the daily sales record will indicate if the cash amount is over or short. For example, if the operation reports gross sales for the day of $2450 and total revenue of $2500, the report will be over by $50 because the total revenue was *more* than the gross sales.

$2500 – $2450 = **$50**

However, if the operation reports gross sales for the day of $3575 and a total revenue of $3550, the report will be short by $25 because the total revenue was *less* than the gross sales.

$3550 – $3575 = **–$25 or ($25) or <25>**

Note: On financial reports negative numbers are shown inside parentheses or angle brackets. For example, "–$25" is written as ($25) or <$25>.

Overages and shortages often occur because the wrong amount of change was returned to a customer or the wrong amount was entered when processing a card payment. Management is responsible for determining the cause of any overages or shortages and taking the appropriate corrective action.

1. Define revenue.

_____

_____

2. Describe the purpose of a guest check.

_____

_____

3. What is a point-of-sale (POS) device?

_____

_____

4. What is the food and beverage subtotal on a guest check that contains the following order: two sandwiches @ $7.95 each, one salad @ $6.95, and two desserts @ $4.50 each?

_____

_____

5. What would be the discount on the order in the previous question if the customers had a coupon good for 15% off their entire food and beverage order?

_____

6. If the food and beverage subtotal on a guest check is $45.50 and the sales tax is 9%, what is the total of the guest check?

_____

7. How much would a 15% gratuity be on a guest check with a food and beverage subtotal of $60.50?

_____

8. If a customer pays with a $50.00 bill and the guest check total is $37.50, how much change should be returned to the customer?

_____

9. If, when completing a daily sales record, the cash at the end of the day is $1240.50 and the cash at the start of the day is $250.25, what is the total cash revenue for the day?

_____

10. On a daily sales record, if a restaurant records gross sales of $3276.65 and a total revenue of $3300.80, would the report be over or short and by how much?

_____

CALCULATING EXPENSES

In addition to calculating revenue, a foodservice operation must calculate its expenses before it can be determined whether the operation is financially successful. There are many expenses involved in building and operating a foodservice operation. These expenses fall under three general categories: capital expenses, cost of goods sold, and operating expenses. **See Figure 7-9.**

Calculating Capital Expenses

When building or improving the physical space occupied by a foodservice operation, the operation will incur capital expenses. A *capital expense* is a cost to an operation for buildings, building improvements, and equipment that is expected to have a useful life longer than one year. Examples of capital expenses include buildings and signage, construction improvements made to a leased space such as new plumbing and electrical systems, major kitchen equipment such as ovens and fryers, and furniture such as tables and chairs.

For example, a business owner decides to lease an empty space in a new strip mall to open a restaurant. If the owner spends $150,000 on construction, $70,000 on equipment, and $35,000 on furniture, the total of those capital expenses would be $255,000.

$$\$150,000 + \$70,000 + \$35,000 = \textbf{\$255,000}$$

It is important to note that the majority of capital expenses are incurred before the foodservice operation is opened for business. This means that the business owner(s) must either borrow or already have the money to pay for these expenses before the business opens since the business cannot start earning money before construction is complete and the business is operational. However, even if an existing operation needs to replace a major piece of equipment after being open for several years, the cost of the replacement equipment is still considered a capital expense.

Calculating Cost of Goods Sold

A popular saying is, "you have to spend money to make money." In accounting terms, this is referred to as the cost of goods sold. In food service, the *cost of goods sold* is the cost of the food and beverage products purchased that are ultimately sold to customers. For example, a bakery must purchase flour, sugar, eggs, milk, and butter before it can make money selling cakes and pastries. Even a can of soda, which is purchased from a beverage supplier and then resold to a customer at a higher price, is considered part of the cost of goods sold. The cost of goods sold is one of the most significant expenses of a foodservice operation.

Expense Categories

Vulcan-Hart, a division of the ITW Food Equipment Group LLC
Capital Expenses

Idaho Potato Commission
Cost of Goods Sold

Operating Expenses

Figure 7-9. The many expenses incurred by a foodservice operation can be broken down into three distinct categories.

When calculating the cost of goods sold, the value of the inventory of food and beverage products already purchased must be taken into account. *Inventory* is the amount of food and beverage products that have been purchased and are currently being stored for future use. For example, a case of pasta stored in a restaurant's dry storage room is considered part of the inventory until it is removed from storage and cooked for customers.

The cost of goods sold is always calculated based on a specific period of time (typically a week or a month). In order to calculate the cost of goods sold during a specific period, the value of inventory must be determined at both the beginning and end of the period. Inventory value is determined by counting the food and beverage items in storage and multiplying the number of units of each item by their as-purchased (AP) unit costs. **See Figure 7-10.** The formula for calculating the inventory value of an individual item is as follows:

True FoodService Equipment, Inc.

$$IV = NU \times APU$$

where
IV = inventory value
NU = number of units
APU = AP Unit Cost

$$\frac{\text{Inventory}}{\text{Value}} = \frac{\text{Number}}{\text{of Units}} \times \text{AP Unit Cost}$$

Calculating Inventory Value

Inventory Sheet—Pizza Restaurant			
Item Type/Name	Number of Units (NU) in Storage	AP Unit Cost (APU)	Inventory Value (IV = NU × APU)
Dry Goods			
All-Purpose Flour	200 lb	$0.40/lb	$ 80.00
Dry Yeast	10 lb	$2.20/lb	$22.00
Granulated Sugar	20 lb	$0.60/lb	$12.00
Salt	40 lb	$0.25/lb	$10.00
Vegetable Oil	15 gal.	$3.85/gal.	$57.75
Pizza Sauce	40 qt	$2.85/qt	$114.00
Meats and Dairy			
Italian Sausage	24 lb	$2.55/lb	$61.20
Pepperoni	18 lb	$3.45/lb	$62.10
Mozzarella Cheese	39 lb	$2.65/lb	$103.35
Fresh Produce			
Yellow Onions	40 lb	$0.39/lb	$15.60
Green Peppers	24 lb	$0.78/lb	$18.72
Mushrooms	16 lb	$1.45/lb	$23.20
Plum Tomatoes	30 lb	$1.88/lb	$56.40
Beverages			
Cola	50 cans	$0.30/can	$15.00
Diet Cola	60 cans	$0.30/can	$18.00
Root Beer	24 bottles	$0.55/bottle	$13.20
Spring Water	38 bottles	$0.40/bottle	$15.20
		Total	**$697.72**

Figure 7-10. The value of an item in inventory is equal to the number of units of the item in storage multiplied by the as-purchased (AP) unit cost of the item.

For example, the inventory value of 200 pounds of all-purpose flour with an AP unit cost of $0.40 per pound is calculated as follows:

IV = NU × APU

$IV = 200 \text{ lb} \times \$0.40/\text{lb}$

$IV = \textbf{\$80.00}$

The inventory value at the beginning of a period plus the amount of food and beverage purchases made during that period represent the total amount of food and beverages that were *available* for sale during that period. By subtracting the inventory value at the end of that period, the *actual* value of the products that were sold (the cost of goods sold) can be calculated. The formula for calculating the cost of goods sold is as follows:

CGS = BIV + FBP – EIV

where

CGS = cost of goods sold
BIV = beginning inventory value
FBP = food and beverage purchases
EIV = ending inventory value

Cost of Good Sold =

Beginning Inventory Value	+	Food and Beverage Purchases	–	Ending Inventory Value

For example, at the beginning of January, a restaurant has a beginning inventory value of $4000. The restaurant then makes a total of $25,000 in food and beverages purchases over the course of the month. If, at the end of January, the restaurant has an ending inventory value of $3000, the cost of goods sold for the month can be calculated as follows:

CGS = BIV + FBP – EIV

$CGS = \$4000 + \$25,000 - \$3000$

$CGS = \textbf{\$26,000}$

In this example, the cost of goods sold was actually higher than the amount of food and beverages purchases made during the month because the restaurant's ending inventory value was lower than its beginning inventory value.

However, if the ending inventory value is higher than the beginning inventory value, the cost of goods sold will be less than the amount of food and beverage purchase made during the period. For example, in June a cafeteria has a beginning inventory value of $15,000. The cafeteria made a total of $80,000 in food and beverage purchases during the month. If, at the end of June, the cafeteria has an ending inventory value of $20,000, the cost of goods sold for the month can be calculated as follows:

CGS = BIV + FBP – EIV

$CGS = \$15,000 + \$80,000 - \$20,000$

$CGS = \textbf{\$75,000}$

When the cost of goods sold is greater than (>) the amount of food and beverage purchases it indicates that the amount of inventory was reduced during the period. Likewise, when the cost of goods sold is less than (<) the amount of food and beverage purchases it indicates that inventory was increased during the period. **See Figure 7-11.**

Calculating Operating Expenses

The final category of expenses in a foodservice operation is operating expenses. An *operating expense* is any ordinary and necessary cost incurred by an operation as a result of carrying out day-to-day operations. Basically, operating expenses are all of the other expenses incurred by an operation that are not capital expenses or part of the cost of goods sold. Some of the major operating expenses of a foodservice operation include payroll expenses, Social Security and unemployment taxes, rent, interest, utilities, insurance and licenses, and supplies.

Payroll Expenses. Payroll expenses are frequently the most significant expense of a foodservice operation. A *payroll expense* is the expense of an operation that includes any money paid to an employee who performs work for the operation. Employees are paid on either a salaried or hourly basis. A salaried employee is paid based on a fixed amount of money. An hourly employee is paid based on an hourly wage multiplied by the number of hours worked.

Employees are further categorized as either exempt or nonexempt as defined by the Fair Labor Standards Act. An exempt employee is not paid an increased wage for working overtime (more than 40 hours in one week). Employees such as front-of-the-house managers and chefs are usually paid a salary and are exempt (not eligible for overtime pay). A nonexempt employee is an employee who is paid an increased wage for working overtime. Line cooks, prep cooks, servers, and dishwashers are usually paid hourly and are nonexempt (eligible for overtime pay).

Hourly, nonexempt employees are required to be paid at least 1½ (1.5) times their regular hourly wage for all hours over 40 worked in a week. For example, a cook earning $10 per hour is paid an overtime wage of $15.00 per hour ($10.00 × 1.5 = $15.00). This overtime wage is commonly referred to as "time-and-a-half."

Payroll expenses for a foodservice operation are based on the gross pay of its employees. *Gross pay* is the total amount of an employee's pay before any deductions are made. Examples of deductions include federal, state, and local income taxes, Social Security taxes, employee-paid health insurance premiums, and contributions to retirement or savings plans. The gross pay for a salaried employee is the same for every pay period. The gross pay for an hourly employee may vary each pay period depending on the number of hours worked.

Figure 7-11. Inventory is reduced when the cost of goods sold is greater than food and beverage purchases and increased when the cost of goods sold is less than food and beverage purchases.

To calculate the gross pay of an hourly employee, regular pay is added to any overtime pay. Regular pay is equal to the number of hours worked up to 40 hours each week multiplied by the regular hourly wage. Overtime pay is equal to any hours worked over 40 hours in a week multiplied by the overtime hourly wage. For example, the gross pay for a line cook who earns $12 per hour and works 50 hours in one week would be calculated as follows:

Regular Pay = 40 hr × $12.00/hr = $480.00

Overtime Pay = 10 hr × ($12.00/hr × 1.5) =
= 10 hr × $18.00
= $180.00/hr

Total Gross Pay = $480.00 + $180.00 = **$660.00**

To calculate the total payroll expenses for a foodservice operation, the gross pay of each employee is added together. **See Figure 7-12.**

Calculating Payroll Expenses

Payroll Summary Report—Week of Jan 1-7, 20XX						
Employee	Regular Hours	Overtime Hours	Pay Rate	Regular Pay	Overtime Pay	Gross Pay
Prep Cook	30	0	$9/hr	$270.00	$0.00	$270.00
Salad Cook	40	0	$10/hr	$400.00	$0.00	$400.00
Line Cook 1	40	10	$12/hr	$480.00	$180.00	$660.00
Line Cook 2	40	5	$12/hr	$480.00	$90.00	$570.00
Server 1	40	0	$6/hr	$240.00	$0.00	$240.00
Server 2	40	6	$6/hr	$240.00	$54.00	$294.00
Chef	—	—	$800/wk	$800.00	—	$800.00
General Manager	—	—	$1000/wk	$1000.00	—	$1000.00
					Total	$4234.00

Figure 7-12. The total payroll expenses of a foodservice operation is equal to the sum of the gross pay for all employees.

The net pay of each employee's paycheck will be less than the employee's gross pay. *Net pay* is the actual amount on an employee's paycheck and is equal to the gross pay minus payroll deductions. One of the most common categories of deductions is taxes.

All employees are subject to pay Social Security and Medicare taxes. Most employees are required to pay federal income taxes, and depending on the city where the job is located, some employees are required to pay state and/or local income taxes. Although taxes are not part of the net pay, they are part of the employee's gross pay and are also considered foodservice operation expenses. The foodservice operation is required to withhold the amount of taxes owed by the employee and then pay those taxes directly to the government on behalf of the employee. **See Figure 7-13.**

Gross Pay vs. Net Pay

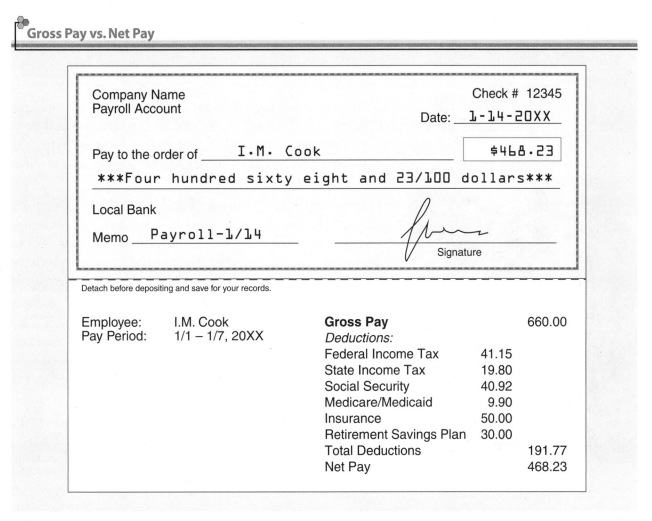

Figure 7-13. The difference between gross pay and net pay is equal to the sum of the payroll deductions.

Social Security and Unemployment Taxes. A foodservice operation is also required to pay additional taxes related to payroll that are above and beyond the taxes included in the gross pay. These taxes include the employer's share of Social Security and unemployment taxes.

Rent. Many foodservice operations lease space for a defined period of time and pay the owner of the building (landlord) rent. Rent is typically calculated based on a dollar amount per square foot. For example, a restaurant renting a 3000 square foot space for $25 per square foot would pay $75,000 in rent.

$$3000 \text{ sq ft} \times \$25/\text{sq ft} = \textbf{\$75,000}$$

Rent is calculated on a yearly basis but is paid on a monthly basis. The monthly payment for the restaurant in this example would be $6250.

$$\$75,000/\text{yr} \times 1 \text{ yr}/12 \text{ months} = \$75,000/12 \text{ months} = \textbf{\$6250/month}$$

Interest. Businesses owners often borrow money to get started, expand, or buy equipment. The interest paid on business loans is an expense. *Interest* is the fee charged by a bank for providing a loan to the borrower.

Utilities. Utility expenses incurred by foodservice operations can include electricity, natural gas, water and sewer service, and telephone service. Computer and media-related services such as the Internet, cable or satellite television, and music services are also included.

Insurance and Licenses. Foodservice operations are required to carry various types of insurance such as liability insurance and workers compensation insurance. In addition to the fees for licenses to sell food, foodservice operations that serve alcoholic beverages are also required to pay for additional licenses and insurance.

Supplies. Supplies are all of the other items besides food and beverages that a foodservice operation purchases in the course of carrying out day-to-day operations such as carryout food containers, sanitation supplies, and office supplies.

Miscellaneous Expenses. Minor expenses that may be applicable to a specific foodservice operation, such as linen rentals, printing charges, and uniforms, are often combined as miscellaneous operating expenses.

Variable Expenses Versus Fixed Expenses

Expenses other than capital expenses can be further categorized as variable expenses or fixed expenses. A *variable expense* is an expense that increases or decreases based on the amount of sales. A *fixed expense* is an expense that does not vary based on the amount of sales. **See Figure 7-14.**

The cost of goods sold is a perfect example of a variable expense because there is a direct relationship between the amount of food and beverages an operation sells and the amount of food and beverage ingredients that it purchases. Other variable expenses include the cost of linen rentals or certain supplies such as carryout food containers.

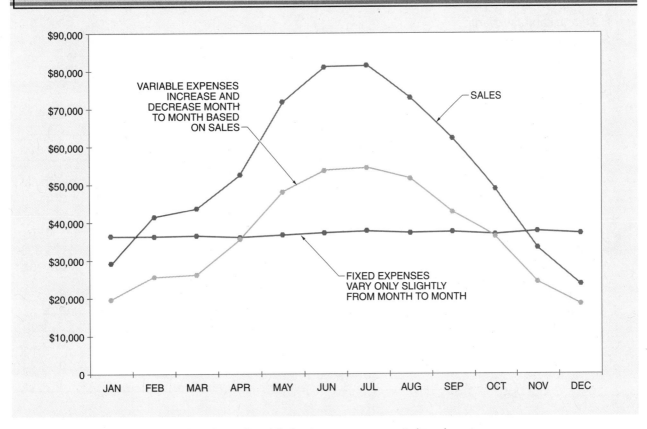

VARIABLE EXPENSES
INCREASE AND
DECREASE MONTH
TO MONTH BASED
ON SALES

SALES

FIXED EXPENSES
VARY ONLY SLIGHTLY
FROM MONTH TO MONTH

Figure 7-14. Variable expenses vary based on sales while fixed expenses are not tied to sales.

Rent and insurance payments are examples of fixed expenses because the same amount is paid every month regardless of the amount of sales. However, a fixed expense is not necessarily the same amount every month. For example, utilities, such as electricity and natural gas, are considered fixed expenses even though the costs vary from month to month. However, those cost variants are not directly tied to the amount of sales of the operation. They are most often due to seasonal changes in the weather such as higher heating expenses in the winter and higher cooling expenses in the summer.

Payroll expenses can be fixed or variable depending on how employees are paid. Generally, the payroll expenses of salaried employees are considered fixed expenses and the payroll expenses of hourly employees are considered variable expenses. When additional food and beverages need to be prepared, which increases sales, the need for additional hourly cooks and servers increases accordingly.

1. Define capital expense.

_____

_____

2. Define cost of goods sold.

_____

_____

3. What is the inventory value of a 50-pound case of whole chickens if the AP unit cost of whole chickens is $1.15 per pound?

_____

4. If a banquet hall has a beginning inventory value in July of $20,500, makes purchases of $125,000 in food and beverage during the month, and has an ending inventory value of $35,000 at the end of July, what were the cost of goods sold in July?

_____

_____

5. Define operating expense.

_____

_____

6. What is the gross pay for the week of an hourly nonexempt dishwasher who earns $9.00 per hour and works 52 hours in that week?

_____

7. Define net pay.

_____

_____

8. What is the monthly rent on a 4000 square foot restaurant if the rent is based on $20 per square foot per year?

_____

_____

9. List three utilities that could be considered operating expenses in a foodservice operation.

_____

10. Explain the difference between a variable expense and a fixed expense.

_____

_____

Quick Quiz® Chapter 7

Flash Cards

Revenue is the amount of money received by a foodservice operation from sales to customers. Guest checks are used as a tool to document the items sold to customers, the prices of those items, and the total amount of money due. Foodservice workers need to know how to perform calculations involving discounts, sales taxes, and gratuities to process guest checks. Point-of-sale (POS) devices and systems are commonly used to process guest checks and also to collect and generate information that helps evaluate the operation's finances. Controls and documentation must be set in place to ensure that the actual amount of money received from customers is equal to the amount of sales documented on guest checks each business day.

It is also important to understand the different types of expenses that are incurred by a foodservice operation. Expenses can be classified as capital expenses related to buildings and equipment, the cost of goods sold related to the purchase of food and beverage ingredients, and operating expenses that are incurred in carrying out day-to-day operations. Payroll expenses are the most significant expenses of a foodservice operation. Calculating payroll involves understanding the different classifications of employees. Some expenses vary based on the amount of sales while others remain relatively constant.

Checkpoint Answers

Checkpoint 7-1

1. Revenue is the total amount of money received by a foodservice operation from sales to customers.

2. A guest check is a form that lists the items sold to a customer, the prices charged for those items, and the total amount of money owed.

3. A point-of-sale (POS) device is an electronic tool used to help process customer orders, print guest checks, track customer and financial information, and generate financial reports.

4. $31.85   [(2 × $7.95) + 6.95 + (2 × $4.50) = $31.85]

5. $4.78   ($31.85 × 15% = $31.85 × 0.15 = $4.78)

6. $49.60   ($45.50 × 9% = $45.50 × 0.09 = $4.10, and $45.50 + $4.10 = $49.60)

7. $9.08   ($60.50 × 15% = 60.50 × 0.15 = $9.08)

8. $12.50   ($50.00 – $37.50 = $12.50)

9. $990.25   ($1240.50 - $250.25 = $990.25)

10. Over $24.15   ($3300.80 - $3276.65 = $24.15)

Checkpoint 7-2

1. A capital expense is a cost to an operation for buildings, building improvements, and equipment that is expected to have a useful life longer than one year.

2. The cost of goods sold is the value of the ingredients used to prepare the food and beverage products sold by a foodservice operation over a specific period of time.

3. $57.50   (50 lb × $1.15/lb = $57.50)

4. $110,500   ($20,500 + $125,000 – $35,000 = $110,500)

5. An operating expense is any ordinary and necessary cost incurred by an operation as a result of carrying out day-to-day operations.

6. $522   (40 hr × $9.00/hr = $360, 12 hr × ($9/hr × 1.5) = 12 hr × $13.50/hr = $162, and $360 + $162 = $522)

7. Net pay is the actual amount on an employee's paycheck and is equal to the gross pay minus payroll deductions.

8. $6666.67/month   (4000 sq ft × $20/sq ft = $80,000, and $80,000/yr × 1 yr/12 months = $6666.67/month)

9. Any of the following are considered operating expenses: electricity, natural gas, water and sewer service, telephone service, internet service, and cable or satellite television, or music services.

10. A variable expense varies based on the sales of an operation and a fixed expense does not vary based on sales.

# Chapter 8

Analyzing Profit and Loss

T he quality of the food and service offered by a foodservice operation is a major factor in determining whether the operation will be successful. However, taking steps to ensure that the operation generates sufficient revenue while keeping expenses to a minimum is just as important. Applying the required math skills is necessary in order to analyze these business aspects of a foodservice operation. Banquets and catered events require additional planning to ensure that all costs are identified and factored into a pricing strategy.

Chapter Objectives

1. Explain the difference between profit and loss.

2. Calculate percent increase of revenue and percent decrease of revenue.

3. List the ways a foodservice operation can minimize expenses.

4. Explain the purpose of a purchase specification.

5. Calculate gross profit and net profit.

6. Interpret a standard profit and loss statement.

7. Interpret a pie chart.

8. Calculate expenses and net profit as a percent of revenue.

9. Estimate the break-even point for a foodservice operation.

10. List topics that are discussed with a customer when estimating the cost of a special event.

11. Use the food cost percentage and contribution margin pricing methods to price special events.

Key Terms

- profit
- loss
- loss leader
- purchase specification
- par stock
- perishable food
- first in, first out (FIFO)
- portion control
- gross profit
- net profit
- standard profit and loss statement
- pie chart
- estimate
- break-even point

National Chicken Council

MAKING A PROFIT

Making a profit in the foodservice industry is a significant challenge. However, determining if a particular foodservice operation is profitable is as simple as comparing revenue to expenses. A *profit* is the amount of money earned by an operation when revenue is greater than (>) expenses. The opposite of a profit is a loss. A *loss* is the amount of money lost by an operation when revenue is less than (<) expenses. These definitions translate into a fundamental math formula that applies to any business.

$$P(L) = TR - TE$$

where
$P(L)$ = profit (loss)
TR = total revenue
TE = total expenses

Profit (Loss) = Total Revenue − Total Expenses

Although making a profit is certainly more likely when an operation offers quality food and service, these characteristics alone do not guarantee financial success. An equal amount of attention must also be paid to the business aspects of running a foodservice operation. In order to be profitable, management and workers need to make consistent efforts to maximize revenue, minimize expenses, and maintain quality standards. **See Figure 8-1.**

Profit and Loss

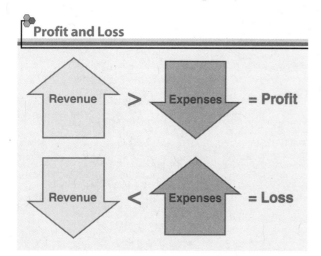

Figure 8-1. A foodservice operation earns a profit if revenue is greater than expenses but experiences a loss if revenue is less than expenses.

Maximizing Revenue

Foodservice operations that consistently offer quality food and service usually see an increase in revenue over time because their customers tell their friends good things about the operation. This "word-of-mouth" advertising generates more customers and, in turn, more revenue for the operation.

Some foodservice operations raise prices to try to increase revenue. Raising prices is sometimes necessary to keep up with increasing expenses. However, higher prices can result in a loss of customers. In contrast, revenue can be maximized by offering enticing menu items, implementing creative marketing and advertising strategies, and training servers to be very effective salespeople.

Sometimes the price of a particular menu item is intentionally set very low. The price can be set so low that the foodservice operation does not make any profit or experiences a small loss on each sale. A menu item priced in this manner is referred to as a loss leader. A *loss leader* is a menu item priced artificially low in order to attract more customers.

For example, a restaurant may advertise "$0.25 buffalo wings from 4:00–6:00 pm." If the buffalo wings cost $0.30 each to prepare, the restaurant will lose $0.05 on each wing sold. However, if the low price attracts a substantial number of additional customers, the profit made on selling other food items and beverages can more than make up for the loss on the buffalo wings.

National Chicken Council

Revenue Goals. Foodservice operations often set goals to increase revenue by a certain percent from year to year. To calculate a new revenue amount based on a percent increase, the following formula is used:

$$NR = OR \times (100\% + \%I)$$

where
NR = new revenue
OR = original revenue
%I = percent increase

$$\text{New Revenue} = \text{Original Revenue} \times \left(100\% + \text{Percent Increase}\right)$$

For example, a foodservice operation that has revenue of $100,000 in the first year and a percent increase of 8% would have a revenue of $108,000 in the second year.

$$NR = OR \times (100\% + \%I)$$

$NR = \$100,000 \times (100\% + 8\%)$

$NR = \$100,000 \times (1.00 + 0.08)$

$NR = \$100,000 \times (1.08)$

$NR = \mathbf{\$108,000}$

If the operation's revenue increased 8% again from the second year to the third year, there would be a third-year revenue of $116,640.

$$NR = OR \times (100\% + \%I)$$

$NR = \$108,000 \times (100\% + 8\%)$

$NR = \$108,000 \times (1.00 + 0.08)$

$NR = \$108,000 \times (1.08)$

$NR = \mathbf{\$116,640}$

Likewise, if the same foodservice operation is successful in increasing revenue by 8% every year for 10 years, its total revenue over 10 years would be almost $1,500,000. In contrast, if the revenue remained constant for 10 years, total revenue would only be $1,000,000. **See Figure 8-2.** Clearly, a small percent increase in revenue from year to year results in a significant increase over a longer period of time.

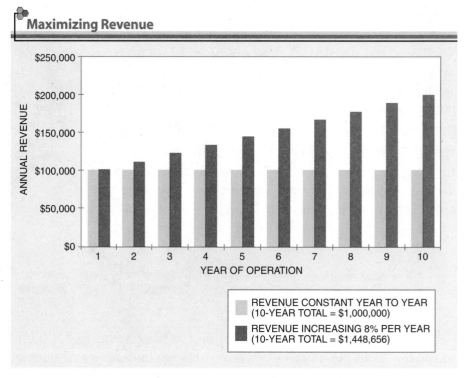

Figure 8-2. Foodservice operations often set goals to increase revenue by a certain percent each year.

Calculating Increases. When actual revenue amounts are known for two periods of time, the percent increase from one period to the next can be calculated. Percent increase can be calculated using the following formula:

$$\%I = (NR - OR) \div OR$$

where
$\%I$ = percent increase
NR = new revenue
OR = original revenue

$$Percent\ Increase = \frac{New\ Revenue - Original\ Revenue}{Original\ Revenue}$$

For example, if a foodservice operation has a first-year revenue of $450,000 and a second-year revenue of $500,000, then the percent increase from the first year to the second year would be 11.1%.

$$\%I = (NR - OR) \div OR$$

$\%I = (\$500,000 - \$450,000) \div \$450,000$

$\%I = \$50,000 \div \$450,000$

$\%I = 0.111$

$\%I = \mathbf{11.1\%}$

Calculating Decreases. If the new revenue is less than the original revenue, the result will be a negative number. This indicates that there was a percent decrease. For example, if a foodservice operation has revenue of $50,000 in June and $48,000 in July, the percent increase would be –4.0%. In this case, revenue actually decreased by 4.0% from June to July.

$$\%I = (NR - OR) \div OR$$

$$\%I = (\$48{,}000 - \$50{,}000) \div \$50{,}000$$

$$\%I = -\$2000 \div \$50{,}000$$

$$\%I = -0.04$$

$$\%I = \textbf{–4.0\% or (4.0\%) or <4.0\%>}$$

It is not uncommon for revenues to fluctuate from month to month. For example, many restaurants are busier during certain times of the year due to changes in seasons. However, revenues should increase each year if a business is to continue to be successful.

Minimizing Expenses

Even though maximizing revenue will contribute to higher profits, it is equally important to minimize expenses. Calculating expenses is a straightforward process for any foodservice operation. However, controlling and minimizing expenses is particularly challenging. The first step in the process is to realize that everything has value. **See Figure 8-3.**

Everything Has Value

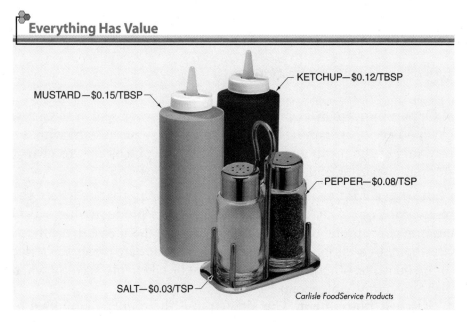

MUSTARD—$0.15/TBSP

KETCHUP—$0.12/TBSP

PEPPER—$0.08/TSP

SALT—$0.03/TSP

Carlisle FoodService Products

Figure 8-3. Even the most minor food products have value and represent an expense to the operation.

Expenses that may seem insignificant at first can add up to substantial amounts over time. For example, when calculating the as-served (AS) cost of a cup of soup, the cost of the ingredients is considered in order to determine an appropriate price to charge customers. However, if a restaurant serves a pack of crackers with every cup of soup, the cost of the crackers must also be considered. The foodservice operation that ignores this additional cost is essentially giving the crackers away.

If a restaurant gave away a $0.05 pack of crackers to 2000 customers per week, the cost would add up to $100 per week and $5200 per year. If this practice continued, the owner would lose thousands of dollars just by overlooking the cost of a $0.05 pack of crackers. However, this cost can be taken into account by increasing the price of the soup by $0.15 to actually make a $0.10 profit on each pack of crackers. Therefore, the owner could earn an additional profit of $10,400 per year instead of a loss of $5200 per year. **See Figure 8-4.**

Turning a Loss into a Profit

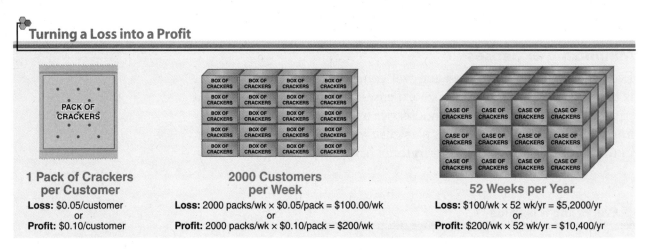

1 Pack of Crackers per Customer
Loss: $0.05/customer
or
Profit: $0.10/customer

2000 Customers per Week
Loss: 2000 packs/wk × $0.05/pack = $100.00/wk
or
Profit: 2000 packs/wk × $0.10/pack = $200/wk

52 Weeks per Year
Loss: $100/wk × 52 wk/yr = $5,2000/yr
or
Profit: $200/wk × 52 wk/yr = $10,400/yr

Figure 8-4. A small profit on a per person basis can add up to a large profit over time.

Opportunities to make money or lose money are present in all aspects of a foodservice operation. The focus on earning a profit starts with the planning of the menu and continues with the purchasing, receiving, storing, preparing, and serving of food.

Planning a Menu. The menu is a document that markets the foodservice operation to the customer. It lists products that satisfy customers, are profitable, and can compete with the alternatives offered elsewhere. In addition to calculating the appropriate prices to charge for menu items, other considerations must be taken into account when planning a profitable menu.

- **Kitchen Equipment.** The kitchen must be properly equipped to efficiently prepare menu items. For example, a kitchen that is not equipped with a deep fryer should not have a large number of deep fried items on the menu. When the appropriate equipment is not available, the operation is less efficient. This inefficiency results in higher payroll expenses.

- **Employee Skill Level.** The techniques used to prepare the items on a menu should be in line with the abilities of the kitchen staff. If employees lack the required skills, food products are likely to be wasted due to errors in preparation causing food expenses to increase. For example, elaborate pastries should not be prepared in-house if a skilled pastry chef is not part of the kitchen staff.

- **Seasonal Ingredients.** Some menu items should only be offered during certain times of the year when key ingredients are readily available at a reasonable price. For example, in the midwestern United States, locally grown sweet corn is widely available at low prices during the summer. However, in winter, the price of corn is much higher and the quality can be lower relative to summer corn.

- **Customer Demand.** All items on a menu should be popular enough with customers that the ingredients required to prepare them are not wasted. For example, a chef may want to offer a dish like antelope or kangaroo, but if the restaurant's customer base tends to shy away from wild game, these special ingredients would go to waste.

Purchasing Food. Purchasing involves the selection and ordering of products required to prepare and serve the items on a menu. A tool used to help ensure that the right products are purchased at the best available price is a purchase specification. A *purchase specification* is a written form listing the specific characteristics of a product that is to be purchased from a supplier. **See Figure 8-5.**

Depending on the product, the information on the purchase specification may include the product's quality or grade, variety or place of origin, size, or packaging requirements. It is also common for the purchase specification to include food safety requirements such as the temperature at which products must be transported and received.

Purchase specifications are prepared and submitted to suppliers to obtain prices. Since they provide the exact parameters that the foodservice operation requires, the operation can be assured that different suppliers are providing prices for the same products.

Purchase Specification—Vegetables

Item Name	Baking potatoes
Variety	Idaho Russet
Grade	US #1
Count per Case	80
Net Weight per Case	50 pounds
Packaging	Heavy-duty cardboard box

Figure 8-5. A purchase specification clearly communicates product specifications required to its suppliers.

For example, if a chef called two different produce suppliers and asked for a price quote on apples, the suppliers might provide prices for two very different types of apples, neither of which may be exactly what the chef wants. However, if each vendor is provided with a purchase specification for US Fancy (highest quality) red delicious apples, averaging 6 ounces each, and packaged in 40-pound cardboard boxes, the chef can be assured that the prices received from each supplier are for the same product.

While a purchase specification is used to help ensure that the appropriate type and quality of product is ordered, a par stock checklist is used to help make sure that the appropriate quantity is ordered. *Par stock* is the amount of a particular product that is to be kept in inventory to ensure that an adequate supply is on hand between deliveries. Par stock values should be set high enough to ensure that the operation does not run out of a product. However, the values should be low enough to avoid disposing of products because they were left in storage too long.

Receiving Food

Detecto, A Division of Cardinal Scale Manufacturing Co.

Figure 8-6. Food products should be checked for quality upon receipt and to make sure the correct amounts were received and the correct prices were charged.

Receiving Food. Another opportunity to help control costs is when food is received from suppliers. All products should be checked upon arrival to ensure that the product received actually matches what was on the purchase specification. Products should be weighed or counted to make sure that the amount or quantity received is the same as specified on the invoice. **See Figure 8-6.**

Special attention should be paid to perishable foods to make sure they are in good condition and at the proper temperature when received. *Perishable food* is food that has a short shelf life and is subject to spoilage and decay. Perishable foods include fresh meats and fish, dairy products, and produce. Perishable items should be purchased frequently in the smallest quantities possible.

Nonperishable items have a much longer shelf life and generally can be kept for six months to a year. These items are normally stored at room temperature in their original packaging. Nonperishable items include canned goods, tea bags, and jars of olives. Since these products have a longer shelf life, they can be purchased in bulk.

Storing Food. Storing foods properly is essential to control costs. Immediately after items have been received and checked-in they should be placed into inventory following the first-in, first-out inventory process. *First-in, first-out (FIFO)* is the process of dating new items as they are placed into inventory and rotating older items to the front of inventory and new items behind. FIFO helps ensure that the oldest items are used first. **See Figure 8-7.**

Preparing Food. Many foods are trimmed or broken down before being served or used in a recipe. Sometimes the parts trimmed from products, such as meats and vegetables, can be used in other recipes, such as stocks or soups, instead of being disposed of as waste. Likewise, high-quality leftovers can potentially be used to prepare other dishes. For example, leftover roasted chicken can be diced for resale as chicken salad.

Serving Food. One of the most important ways foodservice workers can help minimize expenses is to serve food using the proper portion-control techniques. *Portion control* is the process of ensuring that a specific amount of food or beverage is served for a given price. This is accomplished by using portion-control equipment such as ladles, portion-controlled scoops, scales, and slicers. When the quantity of food served is more than the amount the menu price was based upon, food expenses increase and profit is reduced. **See Figure 8-8.**

First-In, First-Out (FIFO) Inventory Process

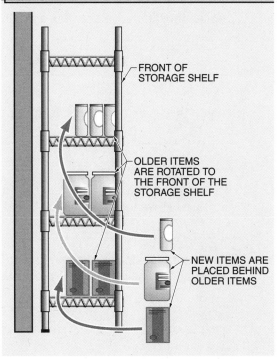

FRONT OF STORAGE SHELF

OLDER ITEMS ARE ROTATED TO THE FRONT OF THE STORAGE SHELF

NEW ITEMS ARE PLACED BEHIND OLDER ITEMS

Figure 8-7. The first-in, first-out (FIFO) inventory process ensures that older products are used before newer products.

Media Clips FIFO Process

Portion-Control Equipment

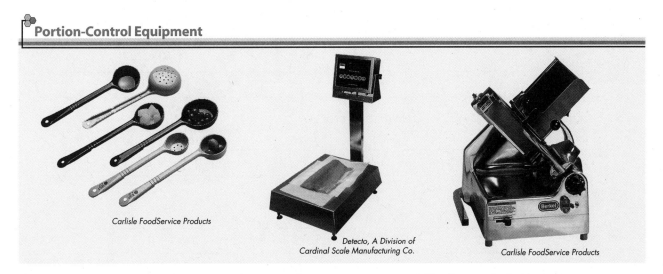

Carlisle FoodService Products

Detecto, A Division of Cardinal Scale Manufacturing Co.

Carlisle FoodService Products

Figure 8-8. Portion-control equipment is used to ensure that a consistent amount of food is served to each customer.

1. Explain the difference between a profit and a loss.

_____

_____

2. If a foodservice operation has a first-year revenue of $250,000 that increases by 9% from the first year to the second year, what is the amount of revenue in the second year?

_____

_____

3. If a restaurant had revenue of $30,000 in September and $33,000 in October, what was the percent increase in revenue from September to October?

_____

4. If a catering company had revenue of $22,000 in January and $21,000 in February, what was the percent decrease in revenue from January to February?

_____

5. List four considerations that should be taken into account when planning a menu.

_____

_____

6. What is a purchase specification?

_____

_____

7. Define par stock.

_____

_____

8. Why should food products be weighed or counted when they are received?

_____

_____

9. Explain the first-in, first-out (FIFO) inventory process.

_____

_____

10. Define portion control.

_____

_____

STANDARD PROFIT AND LOSS

Although every foodservice operation is somewhat unique, the methods for calculating and estimating profits and losses are fairly standard. Foodservice operations ranging from a sandwich shop to a fine-dining restaurant have revenues based on the sales of food and beverages. They also have expenses that can be categorized as either fixed or variable based on the amount of those sales.

Gross Profit Versus Net Profit

There are two categories of profit associated with any foodservice operation: gross profit and net profit. *Gross profit* is the calculated difference between total revenue and the cost of goods sold. *Net profit* is the calculated difference between the gross profit and operating expenses of a foodservice operation.

Calculating Gross Profit. Gross profit can be thought of as the amount of money that is made by the foodservice operation as a result of charging more money for the food items served than the amount spent producing those items. The formula for calculating gross profit is as follows:

$$GP = TR - CGS$$

where
GP = gross profit
TR = total revenue
CGS = cost of goods sold

$$\text{Gross Profit} = \text{Total Revenue} - \text{Cost of Goods Sold}$$

For example, if a restaurant had a total revenue of $31,000 in a month when the cost of goods sold was $10,500, the gross profit would be $20,500.

$$GP = TR - CGS$$

$GP = \$31,000 - \$10,500$

$GP = \textbf{\$20,500}$

Calculating Net Profit. Net profit can be thought of as the amount of money left over after the gross profit is used to pay all the operating expenses incurred by the operation. The formula for calculating net profit is as follows:

$$NP = GP - OE$$

where
NP = net profit
GP = gross profit
OE = operating expenses

$$\text{Net Profit} = \text{Gross Profit} - \text{Operating Expenses}$$

If the restaurant with a gross profit of $20,500 had $18,250 in operating expenses, there would be a net profit of $2250.

$$NP = GP - OE$$

NP = $20,500 – $18, 250

NP = **$2250**

However, if the same restaurant had $21,500 in operating expenses, the restaurant would have a net profit of -$1000, which is a loss of $1000.

$$NP = GP - OE$$

NP = $20,500 – $21,500

NP = **–$1000 or ($1000) or <$1000>**

A loss is not uncommon over certain periods of time. It is, however, especially common in the first several months of a new operation while employees are new and management works to get things operating smoothly.

Creating a Standard Profit and Loss Statement

The relationship between gross profit and net profit is clearly shown on a standard profit and loss statement. A *standard profit and loss statement* is a form that shows the revenue, expenses, and resulting gross and net profit (or loss) over a specific period of time. The purpose of a standard profit and loss statement is to provide management with an indication of the financial status of the foodservice operation.

A standard profit and loss statement can be prepared at a detailed level or a summary level depending on the complexity and needs of the foodservice operation. **See Figure 8-9.** The net profit is the last entry on the statement and is referred to as the "bottom line."

Standard profit and loss statements often show expenses and net profit in terms of a percent of revenue as well as actual dollar amounts. Any expense can be expressed as a percent of revenue by dividing the amount of the expense by the total revenue. For example, if the utility expenses for a restaurant totaled $1125 in a month when total revenue was $34,900, utility expenses for the month would be 3.2% of revenue.

$1125 ÷ $34,900 = 0.032 = **3.2%**

Likewise, if the restaurant's net profit for the month was $3707, then the net profit as a percent of revenue would be 10.6%.

$3707 ÷ $34,900 = 0.106 = **10.6%**

National Chicken Council

Standard Profit and Loss Statement Summary	
Total Revenue	$ 34,908.59
Cost of Goods Sold	$ 10,432.90
Gross Profit	$ 24,475.69
Operating Expenses	$ 20,768.25
Net Profit	$ 3707.44

Standard Profit and Loss Statement Detailed	
Total Revenue	$ 34,908.59
Total Food Revenue	$ 27,801.23
Total Beverage Revenue	$ 7107.36
Cost of Goods Sold	$ 10,432.90
Total Food Costs	$ 7794.38
Total Beverage Costs	$ 2638.52
Gross Profit	$ 24,475.69
Operating Expenses	$ 20,768.25
Payroll Expenses	$ 13,010.15
Social Security Taxes	$ 1050.75
Utilities	$ 1124.67
Rent	$ 2800.00
Interest	$ 325.19
Insurance and Licenses	$ 356.80
Kitchen Supplies	$ 803.21
Sales Tax	$ 1047.26
Miscellaneous	$ 250.22
Net Profit	$ 3707.44

Figure 8-9. A standard profit and loss statement can be prepared at a summary level or a detailed level.

Media Clips Standard Profit and Loss Statement

One way of illustrating expenses and net profit as percentages of revenue is through a pie chart. A *pie chart* is a circle that is divided into pieces where each piece represents a percentage of the whole circle (or pie). The sizes of the pieces are proportional to the percent of the whole pie that each piece represents. For example, if payroll expenses are 37.3% of revenue and rent is 8.0% of revenue, the piece of pie associated with payroll expenses will be over 4 times larger than the piece representing rent. **See Figure 8-10.**

Reviewing financial information in terms of percents instead of dollar values can often be more meaningful. When the percentage associated with a particular expense item changes significantly from one month to another, management should take a close look at that expense to figure out the reason for the change. Sometimes those changes can be good.

For example, if the cost of garbage services as a percent of revenue decreases, it could indicate employees are recycling better. However, changes could also indicate a problem. For example, if the cost of kitchen supplies as a percent of revenue increases significantly, it could indicate that supplies are being wasted. Likewise, the way in which the net profit percentage varies is an indication of how well the operation is doing overall.

Standard Profit and Loss Statement
Detailed

	Dollar Amount	Percent of Revenue
Total Revenue	$ 34,908.59	
Total Food Sales	$ 27,801.23	
Total Beverage Sales	$ 7107.36	
Cost of Goods Sold	$ 10,432.90	29.9%
Total Food Costs	$ 7794.38	
Total Beverage Costs	$ 2638.52	
Gross Profit	$ 24,475.69	
Operating Expenses	$ 20,768.25	
Payroll Expenses	$ 13,010.15	37.3%
Social Security Taxes	$ 1050.75	3.0%
Utilities	$ 1124.67	3.2%
Rent	$ 2800.00	8.0%
Interest	$ 325.19	1.0%
Insurance and Licenses	$ 356.80	1.0%
Kitchen Supplies	$ 803.21	2.3%
Sales Tax	$ 1047.26	3.0%
Miscellaneous	$ 250.22	0.7%
Net Profit	$ 3707.44	10.6%

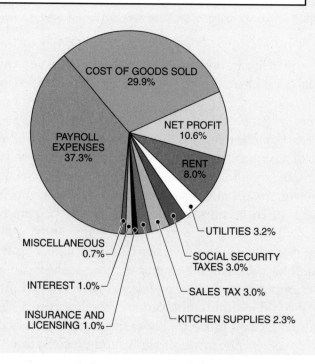

Figure 8-10. Operating expenses and net profit can also be expressed as a percent of revenue.

Estimating Profit and Loss

Creating a standard profit and loss statement is straightforward for a foodservice operation that is up and running. However, it is common to estimate the profit and loss of a foodservice operation before it is operational to help determine whether the new operation will be profitable.

An *estimate* is a calculation that is made using numbers that are approximate and based on the research, experience, and judgment of the person doing the calculation. Since actual numbers for revenue and expenses are not available, certain assumptions need to be made when preparing an estimate.

A good starting point when creating a profit and loss estimate is to perform a break-even analysis to calculate the break-even point of an operation. A *break-even point* is the minimum amount of revenue a foodservice operation must have before the operation can make a profit. The formula for calculating the break-even point for a foodservice operation is as follows:

$BEP = FE \div (100\% - VE\%)$

where

BEP = break-even point

FE = fixed expenses

$VE\%$ = variable expense percentage

$$\text{Break-Even Point} = \frac{\text{Fixed Expenses}}{(100\% - \text{Variable Expense Percentage})}$$

For example, a chef considering opening a restaurant has done a lot of research and estimates that the fixed expenses for rent, utilities, and management-staff payroll will be a total of $24,000 per month. Based on restaurants offering similar food and a similar level of service, the chef estimates that variable expenses will be about 45% of sales. The break-even point for the new restaurant would be estimated as follows:

$BEP = FE \div (100\% - VE\%)$

$BEP = \$24,000/\text{month} \div (100\% - 45\%)$

$BEP = \$24,000/\text{month} \div (1.00 - 0.45)$

$BEP = \$24,000/\text{month} \div 0.55$

$BEP =$ **$43,636/month**

This means that if the restaurant's revenue is more than $43,636 per month, the restaurant will make a profit. However, if the restaurant's revenue is less than $43,636 per month it will experience a loss. **See Figure 8-11.**

Break-Even Point Analysis*

Total Revenue	$30,000	$40,000	$43,636	$50,000	$60,000
Fixed Expenses	$24,000	$24,000	$24,000	$24,000	$24,000
Variable Expenses (45% of Revenue)	$13,500	$18,000	$19,636	$22,500	$27,000
Profit (Loss)	($7500)	($2000)	$0	$3500	$9000

Loss ⟵   **Break Even Point**   ⟶ Profit

* based on fixed expenses of $24,000/month and variable expense percent of 45%

Figure 8-11. A break-even point analysis is a technique used to help analyze the profitability of a new foodservice operation.

New Restaurant Estimate. The break-even analysis can be very helpful in deciding if the restaurant is likely to be profitable. Imagine that the chef has estimated that the restaurant is large enough to serve 100 customers per day (approximately 3000 customers per month). The chef can divide the break-even point by the estimated number of customers to calculate how much each person would be required to spend to generate the required revenue.

$43,636/~~month~~ ÷ 3000 customers/~~month~~ = **$14.54/customer**

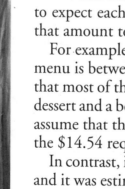

Daniel, NYC

The chef can then decide if $14.54 is a reasonable amount to expect each customer to spend per visit by comparing that amount to the prices of the items on the menu.

For example, the average price for a main course on the menu is between $14.00 and $15.00 and the chef expects that most of the customers will also order an appetizer or a dessert and a beverage. Therefore, it would be reasonable to assume that the average customer would spend more than the $14.54 required to make a profit.

In contrast, if the concept for the restaurant was different and it was estimated that a customer would only spend an average of $9.00, the chef would likely decide that a profit could not be made. In a case such as this, the chef could consider other options such as altering the menu or looking for another location with lower fixed expenses.

Second Location Estimate. Calculating a break-even point is a good place to start when evaluating the profitability of a brand new concept. However, estimates for expanding an existing operation are performed differently.

Consider the owner of a breakfast restaurant who is thinking about opening a second location. The new location will offer the same menu but will be a little larger than the existing location. Since the owner is already established, some of the information from the existing restaurant can be used to evaluate whether the second location will be profitable.

If the menu is going to be the same, the variable expense percent and the amount each customer spends are also likely to be the same. The most significant changes to consider would likely be changes in fixed costs and the number of customers. *Note:* A larger location will likely have higher rent but it also has room for more seats.

In order to estimate the profit of the second location, the owner needs to first estimate the revenue and the expenses. Estimating the revenue is straightforward. The owner expects the average customer to spend $9.80 (the same as the existing location) and has estimated that the new location will serve 4000 customers per month. Therefore, the estimated revenue would be $39,200 per month.

4000 ~~customers~~/month × $9.80/~~customer~~ = **$39,200/month**

The total monthly expenses would be equal to the monthly fixed expenses plus the monthly variable expenses. This owner has already calculated the monthly fixed expenses to be $19,000. The monthly variable expenses are estimated by multiplying the estimated monthly total revenue by the variable expense percentage. The variable expense percentage at the existing restaurant is 40% and is expected to be the same at the new location. The total monthly expenses are estimated using the following formula:

$$TE = FE + (TR \times VE\%)$$

where

TE = total expenses

FE = fixed expenses

TR = total revenue

$VE\%$ = variable expense percentage

$$\text{Total Expenses} = \text{Fixed Expenses} + \left(\text{Total Revenue} \times \text{Variable Expense Percentage} \right)$$

Therefore, the total expenses for the second location are estimated as follows:

$$TE = FE + (TR \times VE\%)$$

TE = $19,000/month + ($39,200/month × 40%)

TE = $19,000/month + ($39,200/month × 0.40)

TE = $19,000/month + $15,680/month

TE = **$34,680/month**

Finally, the monthly profit can be estimated based on the estimated total revenue of $39,200 per month and total expenses of $34,680 per month.

$$P(L) = TR - TE \qquad \textit{Profit (Loss)} = \textit{Total Revenue} - \textit{Total Expenses}$$

$P(L) = \$39,200/\text{month} - \$34,680/\text{month}$

$P(L) = \textbf{\$4520/month}$

Based on this estimate, the owner of the breakfast restaurant can decide if the potential of earning an additional $4520 profit per month justifies the opening of a second location. Since estimates are based on assumptions, there is no guarantee that actual results will match the estimate.

Over 50% of new restaurants fail within the first year of operation. However, by preparing an estimate using good assumptions that are based on thorough research, a new venture is less likely to be among those that fail.

Checkpoint 8-2

Master Math™ Applications

1. Define gross profit.

_____

_____

2. Define net profit.

_____

_____

3. What is the gross profit of a foodservice operation with a total revenue of $28,500 and a cost of goods sold of $9450?

_____

4. What is the net profit of a foodservice operation with a gross profit of $11,090 and operating expenses of $8075?

_____

5. What is a standard profit and loss statement?

_____

_____

6. If a foodservice operation has payroll expenses of $9700 in a month when the total revenue was $32,000, how much were payroll expenses as a percent of revenue?

_____

7. In the following pie chart, which expense is the largest part of total expenses?

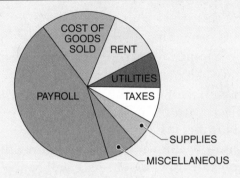

8. What is an estimate?

9. What is the break-even point of a foodservice operation with fixed expenses of $15,000 per month and a variable expense percentage of 38%? Round answer to whole dollars.

10. What is the estimated total expense of a foodservice operation with fixed expenses of $22,000 per month, total revenue of $52,000, and a variable expense percentage of 42%?

SPECIAL EVENT PROFIT AND LOSS

Standard calculations for profit and loss are similar for most ongoing foodservice operations. However, there are some types of food service that present special challenges such as banquets and catered events. Additional topics must be considered when planning and setting prices for special events to ensure the operations are profitable.

Planning Special Events

While there are freestanding banquet facilities, many restaurants also have rooms designated for small banquets. Both types of operations function to accommodate groups of people planning to get together for a special event. Banquet facilities may have multiple rooms that can each accommodate a few hundred to more than 1000 people. Restaurant banquet rooms typically accommodate smaller groups of 20 to 50 people.

On-Site Catering
Food prepared and served at banquet hall or restaurant

Off-Site Catering
Food prepared in professional kitchen and transported to another location

Figure 8-12. Off-site catering differs from banquets (on-site catering) because the food is prepared in one location and transported to another location.

Another term used for "banquet" is "on-site catering" because food is prepared at the same location where it is served. This is not to be confused with off-site catering where food is prepared in a professional kitchen and then transported to a separate location before being served. **See Figure 8-12.**

Because the menu and services vary for both banquets and catered events depending on the specific needs of the customer, a specific price is determined for each event. Determining price is usually the responsibility of a banquet or catering manager. In order to calculate a price that will ensure a profit, the manager will need to discuss the following subjects with the customer.

- **Number of People.** The number of people attending the event must be known in order to figure out how much food to order and prepare, how much space will be required, and the amount of staff needed to prepare and serve the food. Most special event prices are determined on a per person basis.

- **Type of Seating.** The manager will need to know if all guests will be seated at the event and, if so, the number of people seated at each table. Seating arrangements will need to be taken into account when figuring out how much room will be needed for the event. For example, a wedding for 100 guests with 10 guests seated at each table will require less space than if guests are seated 6 to a table. Some events, such as cocktail parties, will not have seating for each guest and require even less space per person.

- **Menu.** The food items that will be served need to be determined in order to calculate the cost of food to be purchased. In addition, it is important to know if all guests will be served the same menu items or if they will be given a choice. If guests are given a choice, the manager will also need to know if guests will be making their choices prior to the event. When decisions are made ahead of time, the exact amount of each item to be prepared is known, which helps keep food expenses to a minimum.

- **Style of Food Service.** The style of service is especially important to know in order to figure out how much staff will be needed to serve the event. For example, more servers will be needed if guests are served plated courses while seated at tables than if guests serve themselves from a buffet. Food costs can be higher for a buffet. This is because when guests are allowed to serve themselves, portions cannot be easily controlled and food is wasted.

- **Style of Beverage Service.** The manager will need to know if beverages will be brought to guests seated at tables or served from a bar. If a bar is used, the manager will need to know if the beverages are to be included in the per-person price of the event (an open bar) or if the guests will be paying for their own beverages (a cash bar).

- **Rentals.** If the banquet facility or caterer does not have all the equipment needed to execute the event, virtually anything can be rented from a rental company. Commonly rented items include tents, tables and chairs, flatware, glassware, linens, chafing dishes, and serving platters. **See Figure 8-13.**
- **Entertainment.** Many events are enhanced by providing entertainment such as bands, musicians, or disk jockeys. Entertainment providers are frequently hired directly by the customer, but the banquet hall or caterer may also make the arrangements. In such cases, the cost of entertainment is built into the total price of the event.

Special Event Rental Items

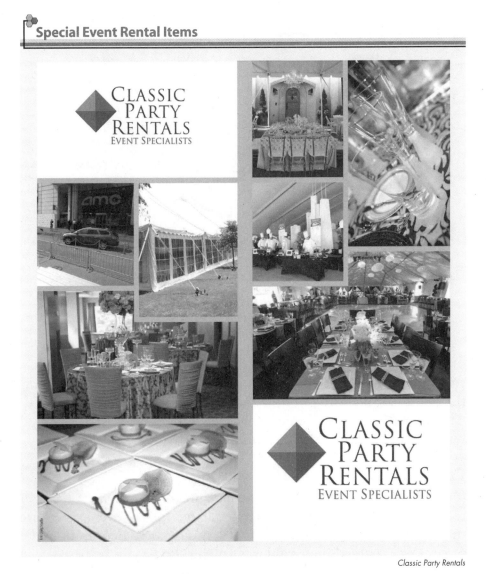

Classic Party Rentals

Figure 8-13. Many foodservice items used for special events are available for rent.

- **Transportation.** For off-site catered events, the method of transporting the food to the service location needs to be determined. Many caterers have their own delivery vehicles. In other circumstances vehicles may need to be rented, delivery services may need to be hired, or customers may simply pick up the food and transport it themselves. Insulated containers are used to keep food at safe holding temperatures during transport. **See Figure 8-14.**

- **Deposits and Method of Payment.** Since many events are planned well in advance, it is common to require the customer to pay a deposit. A *deposit* is an amount of money paid by a customer upon booking a special event to be held at a later date.

Insulated Containers

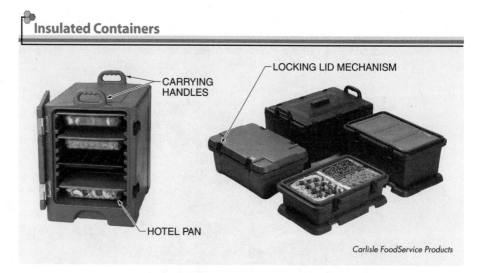

Figure 8-14. Caterers use insulated containers to keep food at proper temperatures during transport.

The deposit is usually based on a percentage of the estimated cost of the event. The manager and customer must have a clear understanding of whether the deposit can be refunded to the customer if the event is cancelled and how the remaining balance will be paid. These details are usually documented in a contract that is signed by both the manager and customer.

Calculating Special Event Profit and Loss

Once the details of an event have been planned, the manager can begin to determine an appropriate price to charge for the event to ensure that a profit is made. The process is very similar to determining the prices to be charged on a restaurant menu. Prices can be based on a target food cost percentage or a contribution margin as discussed in Chapter 6: Calculating Food Costs and Menu Prices.

Food cost percentage pricing is most appropriate for smaller special events that are limited to the serving of food and beverages in a traditional atmosphere (e.g., dinner for a group of people in the party room of a restaurant). For larger special events, especially when temporary staff needs to be hired or outside equipment is rented, contribution margin pricing is more appropriate.

Banquet Profit and Loss. For example, consider a restaurant that has a banquet room that can accommodate a group of up to 50 people. A customer wants to hold a birthday party that will simply include dinner for 50 people with a cash bar. The customer selects a menu of roast beef, baked chicken, mashed potatoes, green beans, tossed salad, and a birthday cake. The customer does not have any requests that require items to be rented or outside staff to be hired.

The AS cost of the menu selected is $8.90 per person and the restaurant has a target food cost percentage of 28%. The target price for the food and beverages would be calculated using the following formula:

TP = ASC ÷ TFC%

TP = $8.90/person ÷ 28%

$$\text{Target Price} = \frac{\text{AS Cost}}{\text{Target Food Cost Percentage}}$$

TP = $8.90/person ÷ 0.28

TP = **$31.79/person**

The restaurant manager would likely adjust this price to $31.95 per person. The restaurant also has a policy that parties of more than 8 people are charged an 18% gratuity. Furthermore, the sales tax rate is 9.5% for that location. The total revenue generated could then be calculated using the following problem solving steps:

Problem-Solving Steps

1. Calculate the food and beverage subtotal based on 50 people at $31.95 per person.

    $31.95/person × 50 people = **$1597.50**

2. Calculate the gratuity.

    $1597.50 × 18% = $1597.50 × 0.18 = **$287.55**

3. Calculate the sales tax.

    $1597.50 × 9.5% = $1597.50 × 0.095 = **$151.76**

4. Calculate the total due (revenue) for the party by adding the food and beverage subtotal, gratuity, and sales tax.

    $1597.50 + $287.55 + 151.76 = **$2036.81**

In order to estimate the gross profit generated by the party, the cost of goods sold must be calculated. For a one-time event, the cost of goods sold is simply equal to the AS cost of the food and beverages since inventory is not maintained for a particular event. Therefore, the cost of goods sold in this example is equal to the AS cost of food per person multiplied by the number of people.

$8.90/person × 50 people = **$445.00**

Then, the gross profit generated from this party can be calculated as follows:

GP = TR − CGS

GP = $2036.81 − $445.00

GP = **$1591.81**

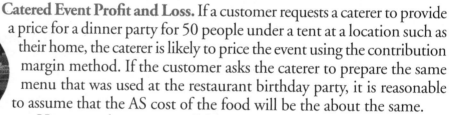

| Gross Profit | = | Total Revenue | − | Cost of Goods Sold |

It is very difficult to determine a net profit from a single event like a party. Many operating costs, such as rent and utilities, apply to the operation as a whole and not to a specific event. However, if the net profit for the restaurant used in the example is typically 9% of total revenue according to the restaurant's profit and loss statements, the party would generate an estimated net profit of $183.31.

$2036.81 × 9% = $2036.81 × 0.09 = **$183.31**

MacArthur Place Hotel, Sonoma

Catered Event Profit and Loss. If a customer requests a caterer to provide a price for a dinner party for 50 people under a tent at a location such as their home, the caterer is likely to price the event using the contribution margin method. If the customer asks the caterer to prepare the same menu that was used at the restaurant birthday party, it is reasonable to assume that the AS cost of the food will be the about the same.

However, the caterer will have other costs to add that did not apply to the restaurant event. **See Figure 8-15.** For example, the tent, as well as tables, chairs, linens, flatware, and glassware will need to be rented. In addition, caterers often hire temporary staff as needed.

If the caterer determines that the total cost to cater the dinner party for 50 people will be $1267.00 and their policy is to charge a contribution margin based on $10.00 per person, the target price for the party would be $35.34 per person.

Target Price = ($1267.00 ÷ 50 people) + $10.00/person

Target Price = $25.34/person + 10.00/person

Target Price = **$35.34/person**

If the caterer adjusts this price to $35.95 per person, adds an 18% gratuity, and has a sales tax rate of 9.5%, the total revenue generated would be calculated using the following problem solving steps:

Problem-Solving Steps

1. Calculate the food and beverage subtotal based on 50 people at $35.95/person.

   $35.95/person × 50 people = **$1797.50**

2. Calculate the gratuity.

   $1797.50 × 18% = $1797.50 × 0.18 = **$323.55**

3. Calculate the sales tax.

   $1797.50 × 9.5% = $1797.50 × 0.095 = **$170.76**

4. Calculate the total due (revenue) for the party by adding the food and beverage subtotal, gratuity, and sales tax.

   $1797.50 + $323.55 + 170.76 = **$2291.81**

Catering Estimate

Catering Estimate

Event Type: Dinner Party
Date: June 1
Customer Name: Joe Customer

Location: Customer's Home
Bar: Provided by Customer
Menu: Roast beef, baked chicken, mashed potatoes, green beans, tossed salad, and birthday cake

Item	Cost per Unit	Number of Units	Total Cost
Food	$8.90/person	50 people	$445.00
Rental Items			
Tent	$200.00	1	$200.00
Tables	$8.50 each	10	$85.00
Chairs	$1.00 each	50	$50.00
Flatware (3 pieces per person)	$0.25 each	150	$37.50
Glassware (2 pieces per person)	$0.30 each	100	$30.00
Plates (2 pieces per person)	$0.50 each	100	$50.00
Linens—Tablecloths	$2.30 each	5	$11.50
Linens—Napkins	$0.20 each	50	$10.00
Payroll Expenses—Employees			
Sous Chef at party to oversee event	$18.00/hr	4 hrs	$72.00
Cooks to prepare food	$12.00/hr	8 hrs	$96.00
Temporary Staff			
Servers (3 @ 4 hours each)	$15.00/hr	12 hrs	$180.00
		Total Cost	**$1267.00**

Figure 8-15. A catering estimate is used to help calculate all of the costs associated with a special event.

As in the banquet example, the cost of goods sold is simply equal to the cost of the food ingredients ($445.00). The gross profit generated by this party is calculated using the following formula:

$$GP = TR - CGS$$

$GP = \$2291.81 - \$445.00$

$GP = \textbf{\$1846.81}$

It is very difficult to calculate a net profit for this particular event as well as for a banquet. However, if the net profit is typically 12% of total revenue according to the caterer's profit and loss statements, the event would generate an estimated net profit of $275.02.

$\$2291.81 \times 12\% = \$2291.81 \times 0.12 = \textbf{\$275.02}$

The key to ensuring that a profit is earned from any special event starts with thorough planning. All costs associated with the event must be considered before providing the customer with a final price that will result in a profit.

Checkpoint 8-3

1. Explain the difference between on-site and off-site catering.

2. When determining the price for a special event, why is it necessary to know the number of people attending?

3. List four things that a caterer may need to rent for a special event.

4. When is it more appropriate to use the method of contribution margin pricing for a special event rather than food cost percentage pricing?

5. If the AS cost of a menu for a banquet is $7.50 per person, what is the target price for the event if the target food cost percentage is 30%?

_____

_____

6. If the price of a special event is $25.00 per person and 80 people attend the event, what is the total revenue generated if a 15% gratuity and 8% sales tax are added to the price?

_____

_____

7. What is the gross profit of a banquet with a total revenue of $4050 and a cost of goods sold of $1025?

_____

_____

8. If the AS cost of a menu for a party is $12.00 and the caterer adds a contribution margin of $9.00 per person, what is the total price if 40 people attend the party?

_____

_____

9. What is the gross profit of a catered event for 200 people if the revenue is $12,500 and the AS cost of the menu is $20 per person?

_____

_____

10. What is the estimated net profit if the total revenue of an event is $5000 and the average net profit as a percent of total revenue is 8%?

_____

_____

Quick Quiz® Chapter 8

Flash Cards

Chapter 8 Summary

A foodservice operation earns a profit when total revenue is more than total expenses. If an operation's total revenue is less than its total expenses, there will be a loss. A foodservice operation is most likely to be successful if steps are taken to maximize revenue, minimize expenses, and offer quality food and service. Care must be taken to minimize expenses throughout the processes of planning the menu and purchasing, receiving, storing, preparing, and serving food.

The gross profit of a foodservice operation is based on the difference between the cost of goods sold and the operation's total revenue. Net profit is based on the difference between the gross profit and operating expenses. Financial information is found on a profit and loss statement. In some cases, it may be necessary to estimate profits and losses.

A key to determining profitability is to calculate the break-even point of a foodservice operation. Ensuring the profitability of special events, such as banquets and catered events, requires a review of the specific needs of the customer.

Key topics to discuss include the number of people attending the event, the menu and style of service, the event location, and items that may need to be rented. This ensures that all of the costs associated with the event are considered before providing the customer with a price so that a profit can be made. Smaller events held in restaurant banquet rooms are typically priced based on a food cost percentage. Larger events, especially when rental items or temporary staff is required, are more likely to be priced based on a contribution margin.

Charlie Trotter's

Checkpoint 8-1

1. A profit occurs when total revenue is more than total expenses and a loss occurs when total revenue is less than total expenses.

2. $272,500 [$250,000 × (1 + 9%) = $250,000 × (1.09) = $272,500]

3. 10% [($33,000 ÷ $30,000) – 1 = 1.10 – 1 = 0.10 = 10.0%]

4. 4.6% [($21,000 ÷ $22,000) –1 = 0.955 –1 = -0.0455 = –4.6%]

5. The following should be considered when planning a menu: kitchen equipment, employee skill level, seasonal availability of ingredients, and customer demand.

6. A purchase specification is a form listing the specific characteristics of a product that is to be purchased.

7. Par stock is the amount of a particular product that is to be kept in inventory to ensure that an adequate supply is available between deliveries.

8. Food products should be weighed or counted when received to make sure the weight or quantity is the same as listed on the invoice.

9. First-in, first-out (FIFO) is the process of dating new items as they are received in inventory and rotating older items to the front while placing new items behind the old.

10. Portion control is the process of ensuring that a specific amount of food or beverage is served for a given price.

Checkpoint 8-2

1. Gross profit is the difference between total revenue and the cost of goods sold of a foodservice operation.

2. Net profit is the difference between the gross profit and the operating expenses of a foodservice operation.

3. $19,050 ($28,500 – $9450 = $19,050)

4. $3015 (11,090 – $8075 = $3015)

5. A profit and loss statement is a form that shows the revenue, expenses, and resulting gross and net profits (or losses) of a foodservice operation over a specific period of time.

6. 30.3 % ($9700 ÷ $32,000 = .303 = 30.3%)

7. Payroll expenses.

continued . . .

8. An estimate is a calculation that is made using numbers that are approximate and based on the research, experience, and judgment of the person doing the calculation.

9. $24,193   [$15,000 ÷ (1 − 38%) = $15,000 ÷ (1− 0.38) = $15,000 ÷ 0.62 = $24,193]

10. $43,840   [$22,000 + ($52,000 × 42%) = $22,000 + ($52,000 × 0.42) = $22,000 + $21,840 = $43,840]

Checkpoint 8-3

1. With on-site catering, the kitchen is located in the same location that the food will be served. With off-site catering, food is prepared in a professional kitchen and then transported to the location where it will be served.

2. The number of people attending a special event must be known in order to calculate the proper amount of food required and the price per person.

3. The following are items commonly rented for catered events: tents, tables, chairs, flatware, glassware, linens, and serving platters.

4. Contribution margin pricing is appropriate when outside staff needs to be hired or equipment needs to be rented.

5. $25.00/person   ($7.50/person ÷ 30% = $7.50/person ÷ 0.3 = $25.00/person)

6. $2460   ($25.00/person × 80 people = $2000; $2000 × 15% = $2000 × 0.15 = $300; $2000 × 8% = $2000 × 0.08 = $160; and $2000 + $300 + $160 = $2460)

7. $3025   ($4050 − $1025 = $3025)

8. $840.00   ($12.00/person + $9.00/person = $21.00/person, and $21.00/person × 40 people = $840.00)

9. $8500   [$12,500 − ($20.00/person × 200 people) = $12,500 − $4000 = $8500]

10. $400   ($5000 × 8% = $5000 × 0.80 = $400)

Appendix

$$\$ \, 7^{\$}_{qt} \, 1 \, 2^{pt} \, 3^{\$}$$
$$5 \times \div \quad \text{TRIM}$$

Math Formulas

Reference Tables

Blank Forms

Math Formulas

Area of a Square or Rectangle

To calculate the area of a two-dimensional square or rectangular object or space, such as a baking sheet or the floor in a banquet room:

$$A = L \times W$$

where
A = area
L = length
W = width

Area = Length × Width

Area of a Circle

To calculate the area of a circle or circular object such as a round cake pan or dining table:

$$A = \pi \times r^2$$

where
A = area
π = 3.14 (constant number)
r = radius

Area = 3.14 × Radius²

Volume of a Rectangular Solid

To calculate the volume or capacity of a three-dimensional rectangular object or space such as a hotel pan or walk-in cooler:

$$V = L \times W \times H$$

where
V = volume
L = length
W = width
H = height (or depth)

Volume = Length × Width × Height

Volume of a Cylinder

To calculate the volume or capacity of cylindrical object such as a stockpot:

$$V = \pi \times r^2 \times H$$

where
V = volume
π = 3.14 (constant number)
r = radius of the base
H = height

Volume = 3.14 × Radius² × Height

Scaling Factor Based on Yield

To calculate the scaling factor when increasing or decreasing a recipe yield:

$$SF = DY \div OY$$

where
SF = scaling factor
DY = desired yield
OY = original yield

$$Scaling\ Factor = \frac{Desired\ Yield}{Original\ Yield}$$

Scaling Factor Based on Product Availability

To calculate the scaling factor when a recipe is scaled based on the available amount of a key ingredient:

$$SF = AA \div OA$$

where
SF = scaling factor
AA = available amount
OA = original amount

$$Scaling\ Factor = \frac{Available\ Amount}{Original\ Amount}$$

New Ingredient Amounts in Scaled Recipes

To calculate the new amount of each ingredient to use when preparing a scaled recipe:

$$NA = OA \times SF$$

where
NA = new amount
OA = original amount
SF = scaling factor

$$New\ Amount = Original\ Amount \times Scaling\ Factor$$

Percentage—Given the Part and the Whole

To calculate the percentage that represents a part of a whole:

$$\% = P \div W$$

where
% = percentage
P = part
W = whole

$$percentage = \frac{part}{whole}$$

The Part—Given the Whole and the Percentage

To calculate the part when the whole and the percentage are known:

$$P = W \times \%$$

where
P = part
W = whole
$\%$ = percentage

$part = whole \times percentage$

The Whole—Given the Percentage and the Part

To calculate the value of the whole when the percentage and the part are known:

$$W = P \div \%$$

where
W = whole
P = part
$\%$ = percentage

$$whole = \frac{part}{percentage}$$

Yield Percentage

To calculate the yield percentage of a food item that is trimmed of waste prior to being served or used in a recipe:

$$YP = EPQ \div APQ$$

where
YP = yield percentage
EPQ = EP quantity
APQ = AP quantity

$$\frac{Yield}{Percentage} = \frac{EP\ Quantity}{AP\ Quantity}$$

As-Purchased Quantity

To calculate the quantity of an ingredient in its as-purchased form that is required to make an edible quantity of food:

$$APQ = EPQ \div YP$$

where
APQ = AP quantity
EPQ = EP quantity
YP = yield percentage

$$\frac{AP}{Quantity} = \frac{EP\ Quantity}{Yield\ Percentage}$$

Edible-Portion Quantity

To calculate the quantity of edible food that can be obtained from a quantity of food in its as-purchased form:

EPQ = APQ × YP

where
EPQ = EP quantity
APQ = AP quantity
YP = yield percentage

$$\text{EP Quantity} = \text{AP Quantity} \times \text{Yield Percentage}$$

Weight of the Main Ingredient in a Formula Based on a Desired Yield

To calculate the weight of the main ingredient in a formula required to obtain a desired yield:

WM = DY ÷ TBP

where

WM = weight of main ingredient
DY = desired yield
TBP = total baker's percentage

$$\text{Weight of Main Ingredient} = \frac{\text{Desired Yield}}{\text{Total Baker's Percentage}}$$

Weight of an Ingredient in a Formula

To calculate the weight of the remaining ingredients in a formula when the weight of the main ingredient is known:

WI = WM × BP

where
WI = weight of ingredient
WM = weight of main ingredient
BP = baker's percentage

$$\text{Weight of Ingredient} = \text{Weight of Main Ingredient} \times \text{Baker's Percentage}$$

Baker's Percentages

To calculate the baker's percentage of each ingredient in a formula when creating a formula from a recipe:

BP = WI ÷ WM

where
BP = baker's percentage
WI = weight of ingredient
WM = weight of main ingredient

$$\text{Baker's Percentage} = \frac{\text{Weight of Ingredient}}{\text{Weight of Main Ingredient}}$$

Math Formulas (continued)

As-Purchased Unit Cost

To calculate the unit cost of a food or beverage item based the as-purchased cost.

$$APU = APC \div NU$$

where
APU = AP unit cost
APC = AP cost
NU = number of units

$$AP\ Unit\ Cost = \frac{AP\ Cost}{Number\ of\ Units}$$

Edible-Portion Unit Cost

To calculate the unit cost of a food item after taking into account the cost of the waste generated by trimming:

$$EPU = APU \div YP$$

where
EPU = EP unit cost
APU = AP unit cost
YP = yield percentage

$$EP\ Unit\ Cost = \frac{AP\ Unit\ Cost}{Yield\ Percentage}$$

Menu-Item Food Cost Percentage

To calculate the percentage that relates the cost of the ingredients used to prepare a menu item to the price charged for a menu item:

$$IFC\% = ASC \div MP$$

where
$IFC\%$ = menu-item food cost percentage
ASC = AS cost
MP = menu price

$$Menu\text{-}Item\ Food\ Cost\ Percentage = \frac{AS\ Cost}{Menu\ Price}$$

Overall Food Cost Percentage

To calculate the percentage that relates a foodservice operation's total cost of food to its total sales of food:

$$OFC\% = FC \div FS$$

where
$OFC\%$ = overall food cost percentage
FC = total food costs
FS = total food sales

$$Overall\ Food\ Cost\ Percentage = \frac{Total\ Food\ Costs}{Total\ Food\ Sales}$$

Food Item Target Price

To calculate the menu price of a food item based on a target food cost percentage:

TP = ASC ÷ TFC%

where
TP = target price
ASC = AS cost of a menu item
TFC% = target food cost percentage

$$\text{Target Price} = \frac{\text{AS Cost}}{\text{Target Food Cost Percentage}}$$

Food Item Menu Price

To calculate the price of a food item on a menu when the as-served cost of the food item and the menu-item food cost percentage are known:

MP = ASC ÷ IFC%

where
MP = menu price
ASC = AS cost
IFC% = menu-item food cost percentage

$$\text{Menu Price} = \frac{\text{AS Cost}}{\text{Menu-Item Food Cost Percentage}}$$

Beverage Item Menu Price

To calculate the price of a beverage item on a menu when the as-served cost of the beverage item and the menu-item beverage cost percentage are known:

IBC% = ASC ÷ MP

where
IBC% = menu-item beverage cost percentage
ASC = AS cost
MP = menu price

$$\text{Menu Item Beverage Cost Percentage} = \frac{\text{AS Cost}}{\text{Menu Price}}$$

Beverage Item Target Price

To calculate the price charged for a beverage item on a menu based on the foodservice operation's target beverage cost percentage.

TP = ASC ÷ TBC%

where
TP = target price
ASC = AS cost
TBC% = target beverage cost percentage

$$\text{Target Price} = \frac{\text{AS Cost}}{\text{Target Beverage Cost Percentage}}$$

Math Formulas (continued)

Overall Beverage Cost Percentage

To calculate the percentage that relates a total beverage costs to total beverage sales:

$$OBC\% = BC \div BS$$

where
OBC% = overall beverage cost percentage
BC = total beverage costs
BS = total beverage sales

$$\text{Overall Beverage Cost Percentage} = \frac{\text{Total Beverage Costs}}{\text{Total Beverage Sales}}$$

Inventory Value

To calculate the dollar value of a product in inventory:

$$IV = NU \times APC$$

where
IV = inventory value
NU = number of units
APU = AP Unit Cost

$$\text{Inventory Value} = \text{Number of Units} \times \text{AP Unit Cost}$$

Cost of Goods Sold

To calculate the cost of the food and beverages that were used to generate revenue over a specific period of time:

$$CGS = BIV + FBP - EIV$$

where

CSG = cost of goods sold
BIV = beginning inventory value
FBP = food and beverage purchases
EIV = ending inventory value

$$\text{Cost of Good Sold} = \text{Beginning Inventory Value} + \text{Food and Beverage Purchases} - \text{Ending Inventory Value}$$

Profit (Loss)

To calculate the amount of money earned or lost by a foodservice operation based on the operation's total revenue and total expenses over a specific period of time:

$$P(L) = TR - TE$$

where
P(L) = profit (loss)
TR = total revenue
TE = total expenses

$$\text{Profit (Loss)} = \text{Total Revenue} - \text{Total Expenses}$$

Revenue Based on a Percent Increase

To calculate the new amount of revenue when the original amount of revenue is increased by a specific percent:

$$NR = OR \times (100\% + \%I)$$

where
NR = new revenue
OR = original revenue
%I = percent increase

$$\text{New Revenue} = \text{Original Revenue} \times \left(100\% + \text{Percent Increase}\right)$$

Percent Increase

To calculate a percent increase based on an original amount of revenue and a new amount of revenue:

$$\%I = (NR - OR) \div OR$$

where
%I = percent increase
NR = new revenue
OR = original revenue

$$\text{Percent Increase} = \frac{\text{New Revenue} - \text{Original Revenue}}{\text{Original Revenue}}$$

Gross Profit

To calculate the amount of money earned by a foodservice operation as a result of charging more money for the food items served than the amount spent producing those items:

$$GP = TR - CGS$$

where
GP = gross profit
TR = total revenue
CGS = cost of goods sold

$$\text{Gross Profit} = \text{Total Revenue} - \text{Cost of Goods Sold}$$

Net Profit

To calculate the amount of money earned by a foodservice operation after taking into account the cost of goods sold and all operating expenses:

$$NP = GP - OE$$

where
NP = net profit
GP = gross profit
OE = operating expenses

$$\text{Net Profit} = \text{Gross Profit} - \text{Operating Expenses}$$

Break-Even Point

To estimate the minimum amount of revenue a foodservice operation must have before the operation can make a profit:

$$BEP = FE \div (1 - VE\%)$$

where
BEP = break-even point
FE = fixed expenses
$VE\%$ = variable expense percentage

$$\text{Break-Even Point} = \frac{\text{Fixed Expenses}}{(100\% - \text{Variable Expense Percentage})}$$

Estimating Total Expenses

To estimate the total expenses of a foodservice operation based on fixed expenses and the estimated variable expense percentage:

$$TE = FE + (TR \times VE\%)$$

where
TE = total expenses
FE = fixed expenses
TR = total revenue
$VE\%$ = variable expense percentage

$$\text{Total Expenses} = \text{Fixed Expenses} + \left(\text{Total Revenue} \times \text{Variable Expense Percentage} \right)$$

Reference Tables

Common Food Service Units of Measure

Volume Units

Customary System

Unit	Abbreviation
teaspoon	tsp
tablespoon	tbsp
fluid ounce	fl oz
cup	c
pint	pt
quart	qt
gallon	gal.

Metric System

Unit	Abbreviation
liter	L
milliliter	mL

Temperature Units

Customary System

Unit	Abbreviation
degrees Fahrenheit	°F

Metric System

Unit	Abbreviation
degrees Celsius	°C

Weight Units

Customary System

Unit	Abbreviation
ounce	oz
pound	lb or #

Metric System

Unit	Abbreviation
milligram	mg
gram	g
kilogram	kg

Distance Units

Customary System

Unit	Abbreviation
inch	in.
foot	ft

Metric System

Unit	Abbreviation
millimeter	mm
centimeter	cm
meter	m

Metric Prefixes

Prefix		Value
kilo (k)	=	1000 ×
hecto (h)	=	100 ×
deka (da)	=	10 ×
deci (d)	=	0.1 ×
centi (c)	=	0.01 ×
milli (m)	=	0.001 ×

Portion-Controlled Scoop Equivalents

Scoop Number	Fluid Ounces per Scoop	Scoop Number	Fluid Ounces per Scoop
4	8	16	2
5	6.4	20	1.6
6	5.33	24	1.33
8	4	30	1.06
10	3.2	40	0.8
12	2.66	60	0.53

Volume Measurement Equivalents

Customary System Equivalents

1 gallon = 4 quarts = 8 pints = 16 cups = 128 fluid ounces = 256 tablespoons = 768 teaspoons

1 quart = 2 pints = 4 cups = 32 fluid ounces = 64 tablespoons = 192 teaspoons

1 pint = 2 cups = 16 fluid ounces = 32 tablespoons = 96 teaspoons

1 cup = 8 fluid ounces = 16 tablespoons = 48 teaspoons

1 fluid ounce = 2 tablespoons = 6 teaspoons

1 tablespoon = 3 teaspoons

Customary–Metric Equivalents

1 gallon (gal.) = 3.79 liters (L)

1 quart (qt) = 0.95 liters (L)

1 cup = 236.6 milliliters (mL)

1 fluid ounce (fl oz) = 29.6 milliliters (mL)

1 teaspoon (tsp) = 5 milliliters (mL)

1 liter (L) = 1.06 quart (qt)

1 liter (L) = 33.8 fluid ounces (fl oz)

Metric System Equivalents

1 kiloliter (kL)	=	1000 liters (L)
1 hectoliter (hL)	=	100 liters (L)
1 dekaliter (daL)	=	10 liters (L)
1 deciliter (dL)	=	0.1 liters (L)
1 centiliter (cL)	=	0.01 liters (L)
1 milliliter (mL)	=	0.001 liters (L)

Weight Measurement Equivalents

Customary System Equivalents
1 pound (lb) = 16 ounces (oz)

Customary–Metric Equivalents
1 pound (lb) = 0.454 killograms (kg)
1 pound (lb) = 454 grams (g)
1 ounce (oz) = 28.4 grams (g)
1 kilogram (kg) = 2.2 pounds (lb)

Metric System Equivalents		
1 kilogram (kg)	=	1000 grams (g)
1 hectogram (hg)	=	100 grams (g)
1 dekagram (dag)	=	10 grams (g)
1 decigram (dg)	=	0.1 grams (g)
1 centigram (cg)	=	0.01 grams (g)
1 milligram (mg)	=	0.001 grams (g)

Volume-to-Weight Equivalents of Common Food Products ...

Item Name	Volume	Ounces	Item Name	Volume	Ounces
Allspice, ground	tbsp	0.25	Brussels sprouts	c	4
Almonds, blanched	c	5	Butter	c	8
Apples			Cabbage, red or green	c	2.5
peeled, ½ in cubes	c	5	Canteloupe melon	c	5.75
pie, canned	c	9	Carrots, raw sliced or diced	c	5
sauce, canned	c	8	Carrots, cooked	c	5.5
Apricots, dried	c	4.5	Casaba melon	c	6
Asparagus spears†	c	6	Cauliflower, florets	c	4.75
Baking powder	tbsp	0.5	Celery, diced	c	4
Baking soda	tbsp	0.5	Celery seed	tbsp	0.25
Bananas, sliced	c	7	Cheese		
Barley, raw	c	8	cubes	c	6
Beans			cottage or cream	c	8
great northern	c	6.5	grated, hard (e.g., parmesan)	c	4
green, trimmed	c	5	grated, medium (e.g., cheddar)	c	4
kidney, dried	c	6	grated, soft (e.g., fresh goat)	c	4.75
kidney, cooked	c	6.75	Cherries		
lima, dried	c	6.75	dried	c	6
navy, dried	c	7	fresh	c	5.75
white beans, small	c	7.5	pitted	c	5.5
Bean sprouts	c	4	Chicken, cooked, cubed	c	5.5
Blueberries			Chili powder	tbsp	¼
canned	c	6.75	Chili sauce	c	11.25
fresh	c	7	Chocolate		
dried	c	5.5	chips	c	6
Breadcrumbs			grated	c	4
cake or pastry			melted	c	8
dried, fine ground	c	4	Cinnamon, ground	tbsp	0.25
fresh	c	2	Cloves		
Broccoli			ground	tbsp	0.25
florets	c	2.75	whole	c	3
spears	c	3	Cocoa	c	3.75

…Volume-to-Weight Equivalents of Common Food Products…

Item Name	Volume	Ounces	Item Name	Volume	Ounces
Coconut			Greens, collard, mustard	c	2
loose shredded	c	2.5	Half and half	c	8.5
packed shredded	c	3.25	Ham, cooked, diced	c	5.33
Coconut milk	c	8.5	Honey	c	12
Corn, fresh, kernels	c	5.5	Honeydew melon	c	6
Corn flakes	c	1	Horseradish	tbsp	0.5
Cornmeal, raw	c	5	Jam	c	12
Cornstarch	c	5.33	Jelly	c	10.5
Corn syrup	c	12	Kale	c	2.5
Cracker crumbs	c	3.5	Kiwi	c	7
Cranberries, raw	c	4	Lard	c	8
Cream			Leeks	c	3.25
whipped	c	4	Lemon juice	c	8
whipping	c	8.5	Lettuce		
Cream of tartar	tbsp	0.33	butter, grean leaf, red leaf	c	2
Crenshaw melon	c	6	iceberg chopped	c	2.25
Cucumbers, diced	c	5.5	romaine	c	2
Currants, dried	c	5.33	shredded	c	2.25
Curry powder	tbsp	0.25	Lime juice (from fresh lime)	c	8.1
Dates, pitted	c	6	Mango	c	6
Eggs			Margarine	c	8
dried whites	c	3.25	Marshmallows, large	10 ea	2
dried yolks	c	3	Mayonnaise	c	8
fresh whites (9)	c	8	Milk		
fresh yolks (10)	c	8	condensed	c	10.75
raw, shelled (5 eggs)	c	8	evaporated	c	9
Extracts	tbsp	0.5	liquid, whole	c	8
Figs	c	7.9	nonfat dry	c	5
Figs, dried, chopped	c	6.5	Molasses	c	12
Flour			Mustard		
all-purpose	c	4.5	dry, ground	c	3.5
cake or pastry	c	4	prepared	tbsp	.5
high-gluten	c	4.75	seed	tbsp	0.33
rye	c	3.5	Noodles, cooked	c	5.33
whole wheat	c	4.5	Nuts	c	4
Garlic	c	5	Nutmeg, ground	tbsp	0.25
Gelatin, granulated	c	5	Oats, rolled	c	3.5
Ginger, ground	tbsp	.2	Oil, vegetable	c	8
Glucose	c	12	Onions		
Graham cracker crumbs	c	4	yellow, diced	c	6
Grapefruit	c	7.5	green (scallions)	c	2
Grapes			Orange juice	c	8.1
whole	c	3.75	Oysters, shucked	c	8
sliced	c	5.75	Papaya	c	5.5

...Volume-to-Weight Equivalents of Common Food Products

Item Name	Volume	Ounces	Item Name	Volume	Ounces
Paprika	tbsp	0.25	Sesame seeds	tbsp	0.33
Parsley, coarsely chopped	c	1.25	Shallots, diced	c	6.4
Peaches			Shortening	c	6.75
pitted, diced	c	8	Sour cream	c	8.5
canned, diced	c	8	Spinach, leaf	c	0.8
Peanut butter	c	8	Spices, ground	tbsp	0.25
Peanuts	c	5	Squash		
Pears, fresh, cored and diced†	c	5.25	acorn	c	5
Peas			butternut	c	5
dry split	c	7	spaghetti	c	5
frozen	c	3.5	summer	c	3.8
snap	c	3	zucchini	c	3.8
snow	c	3	Strawberries	c	5.5
Pecans	c	4	Sweet potatoes, diced	c	5.1
Pepper, ground	tbsp	0.25	Sugar		
Peppers, green and red, chopped	c	5.25	brown, lightly packed	c	5.5
Pimiento, chopped	c	6.75	brown, solidly packed	c	8
Pineapple			granulated	c	8
crushed	c	8	powdered, sifted	c	4.5
diced	c	5	Tapioca, pearl	c	5
Pomegranate	c	4.25	Tea, loose-leaf	c	2.5
Poppy seeds	c	5	Tomatoes		
Potatoes,			canned	c	8
cooked, diced, or mashed	c	7	cored and peeled	c	6
russet, peeled and diced	c	5	fresh, diced	c	6
Prunes, dried	c	6	peeled, seeded, and chopped	c	6
Raisins	c	5.33	plum cored	c	6
Raspberries, fresh	c	5	Tuna	c	8
Rice			Vanilla	tbsp	0.25
cooked	c	8	Vinegar	c	8
uncooked long grain	c	6.75	Walnuts, shelled	c	4
uncooked short grain	c	7	Water	c	8
Rutabagas, cubed	c	4.75	Watermelons	c	5.5
Sage, ground	tbsp	0.125	Wild rice	c	6
Salad dressing	c	8	Yeast		
Salmon, canned	c	8	compressed cake	tbsp	0.33
Salt			envelope active dry	tbsp	0.5
granulated	tbsp	0.5			
kosher	tbsp	0.33			

Approximate Yield Percentages of Fruits and Vegetables

Item Name	Yield %	Item Name	Yield %
Apples, peeled and cored	85	Lime juice (from fresh limes)	45
Asparagus, spears, trimmed	60	Mangoes	65
Bananas	65	Onions, large	93
Cantaloupes	56	Orange juice (from fresh oranges)	35
Casaba melons	59	Papaya	69
Cherries, pitted	89	Parsley	75
Crenshaw melons	65	Peaches, pitted	79
Broccoli, spears	75	Pears, cored	78
Broccoli, florets	65	Peas, snow and snap	95
Cabbage (red and green)	80	Peppers, bell	85
Carrots	85	Pineapples	50
Cauliflower, florets	65	Pomegranates	50
Celery	70	Potatoes, russet	80
Cucumbers	94	Scallions (green onions)	85
Figs	97	Shallots	92
Garlic	85	Spinach, leaf	68
Grapefruit	55	Squash, acorn	78
Grapes, stemmed	95	Squash, butternut	85
Green beans, trimmed	90	Squash, spaghetti	70
Greens, collard, mustard, or turnip	70	Squash, summer	95
Honeydew melons	60	Squash, zucchini	95
Kale	70	Strawberries, stemmed	95
Kiwis	87	Sweet potatoes, diced	90
Leeks	50	Tomatoes, cored	98
Lemon juice (from fresh lemons)	45	Tomatoes, cored and diced	93
Lettuce, green leaf, red leaf, and butter	84	Tomatoes, peeled, seeded, and diced	80
Lettuce, iceberg	75	Tomatoes, plum cored	96
Lettuce, romaine	78	Watermelons	50

Common Professional Kitchen Ratios

Biscuits	3 parts flour : 2 parts liquid : 1 part fat (by weight)
Boiled Rice	1 part uncooked rice : 2 parts liquid (by volume)
Cookie Dough	3 parts flour : 2 parts fat : 1 part sugar (by weight)
Mirepoix	2 parts onions : 1 part carrot : 1 part celery
Pie Dough	3 parts flour : 2 parts fat : 1 part water (by weight)
Roux	1 part flour : 1 part fat (by weight)
Simple Syrup	1 part granulated sugar : 1 part water
Slurry	2 parts liquid : 1 part starch (by volume)
Vinaigrette	3 parts oil : 1 part vinegar (by volume)

Blank Forms

Standardized Recipe

Recipe Name:

Yield: **Cooking Temperature:**

Portion Size: **Cooking Time:**

Amount Ingredients	Procedure

Nutrition info (per serving):

Pricing Form

Menu Item Name:

Number of Portions:

AS Cost per Portion: **Menu Price:**

Target Food Cost %:   **Menu-Item Food Cost %:**

Target Price:

Ingredients	EP Quantity	EP Unit of Measure	AP Unit Cost (or recipe cost)	Yield Percentage	EP Unit Cost	Total Ingredient Cost
					Total Cost	

Daily Sales Record

Total Sales	
Total Food Sales	
Total Beverage Sales	
Sales Tax (%)	
Gross Sales (*Total Sales + Sales Tax*)	
Cash at End of Day	
Cash at Start of Day	
Total Cash Revenue (*Cash at End of Day – Cash at Start of Day*)	
Total Card Revenue	
Credit Cards	
Debit Cards	
Gift Cards	
Total Revenue (*Total Cash Revenue + Total Card Revenue*)	
Over (Short) (*Total Revenue – Gross Sales*)	

Report completed by: _____   _____

                                                       Signature   Date

Glossary

A

addition: The process of combining two or more numbers into a single number to find the sum.

angle: A measurement that indicates the relationship between two lines when the lines intersect one another.

area: The size of a two-dimensional (flat) surface.

as-purchased (AP) cost: The amount paid for a product in the form it was ordered and received.

as-purchased (AP) quantity: The original amount of a food item as it is ordered and received.

as-purchased (AP) unit cost: The unit cost of a food item based on the form in which it is ordered and received.

as-served (AS) cost: The cost of a menu item as it is served to a customer.

average: The sum of a set of numbers divided by how many numbers there are in the set.

B

baker's percentage: The weight of a particular ingredient expressed as a percentage based on the weight of the main ingredient in a formula.

balance scale: A scale with two platforms that uses a counterbalance system to measure weight.

beverage cost percentage: A percentage that indicates how the cost of beverages relates to menu prices and beverage sales of a foodservice operation.

break-even point: The minimum amount of revenue a foodservice operation must have before the operation can make a profit.

C

cancelling: The process of crossing out and eliminating matching units in the numerators and denominators of fractions in a conversion calculation.

capacity: The volume that can be placed inside an empty three-dimensional object.

capital expense: A cost to an operation for buildings, building improvements, and equipment that is expected to have a useful life longer than one year.

common denominator: A denominator that is the same number in two fractions or more.

contribution margin: The amount added to the AS cost of a menu item to determine a menu price.

conversion factor: A constant number that defines the relationship between different units of measure.

converting: The process of changing a measurement with one unit of measure to an equivalent measurement with a different unit of measure.

cooking-loss yield test: A procedure used to determine the yield percentage of a food item that loses weight during the cooking process.

cost of goods sold: The cost of the food and beverage products purchased that are ultimately sold to customers.

data table: A collection of information that is organized into a set of rows and columns.

decimal: A number or set of numbers that represent part of a whole and can be expressed as a fraction with a denominator that is a factor of 10.

denominator: The number in a fraction at the bottom (or to the right) of a fraction bar that represents the number of parts into which a whole is divided.

density: The measure of how much a given volume of a substance weighs.

deposit: An amount of money paid by a customer upon booking a special event to be held at a later date.

diameter: The length of a line drawn through the center of a circle to both edges.

digital scale: An electronic scale with a sensor that measures weight and displays the result electronically.

discount: A percentage or fixed amount by which the subtotal is reduced due to a promotion offered by the foodservice operation.

dividend: The number being divided.

division: The process of counting how many times one number can go into another number.

divisor: The number by which the dividend is divided.

dry measuring cup: A volume measurement tool, shaped like a cup, with a short handle used to measure a specific volume of dry ingredients.

edible-portion (EP) unit cost: The unit cost of a food or beverage item after taking into account the cost of the waste generated by trimming.

edible-portion (EP) quantity: The amount of a food item that remains after trimming and is ready to be served or used in a recipe.

estimate: A calculation that is made using numbers that are approximate and based on the research, experience, and judgment of the person doing the calculation.

first-in, first-out (FIFO): The process of dating new items as they are placed into inventory and rotating older items to the front of inventory and new items behind.

fixed expense: An expense that does not vary based on the amount of sales.

food cost percentage: A percentage that indicates how the cost of food relates to the menu prices and food sales of a foodservice operation.

formula: A recipe format in which all ingredient quantities are provided as baker's percentages.

fraction: A part of a whole number.

G

gratuity: The amount of money left by a customer as thanks for the services rendered.

gross pay: The total amount of an employee's pay before any deductions are made.

gross profit: The calculated difference between total revenue and the cost of goods sold.

I

improper fraction: A fraction in which the numerator is larger than the denominator.

inventory: The amount of food and beverage products that have been purchased and are currently being stored for future use.

L

ladle: A fixed-size cup attached to a long handle.

liquid measuring cup: A volume measurement tool, shaped like a cup or pitcher, with graduated markings on the side that indicate the volume of a liquid.

loss: The amount of money lost by an operation when revenue is less than (<) expenses.

loss leader: A menu item priced artificially low in order to attract more customers.

lowest common denominator: The smallest number into which the denominators of a group of fractions divide evenly.

M

measurement: A number with a corresponding unit of measure.

measurement equivalent: The amount of one unit of measure that is equal to another unit of measure.

measuring spoon: A small volume measurement tool, shaped like a spoon, that is used to measure liquid or dry ingredients.

mechanical scale: A scale with a spring-loaded platform and a mechanical-dial display.

median: The middle value within a set of numbers that are arranged in numerical order.

menu-item food cost percentage: The AS cost of a menu item divided by the menu price, written as a percent.

mill: A term used to represent $\frac{1}{1000}$ of a dollar ($0.001).

mixed number: A combination of a whole number and a fraction.

mode: The value that appears most frequently within a set of numbers.

multiplication: The process of adding one number to itself any number of times to find the product.

multiplicand: The number being multiplied.

multiplier: The number by which the multiplicand is multiplied.

net pay: The actual amount on an employee's paycheck and is equal to the gross pay minus payroll deductions.

net profit: The calculated difference between the gross profit and operating expenses of a foodservice operation.

numerator: The number in a fraction at the top (or to the left) of a fraction bar that represents the parts of a whole.

operating expense: Any ordinary and necessary cost incurred by an operation as a result of carrying out day-to-day operations.

overall food cost percentage: The total amount of money a foodservice operation spends on food divided by the total food sales over a defined period of time, written as a percent.

par stock: The amount of a particular product that is to be kept in inventory to ensure that an adequate supply is on hand between deliveries.

payroll expense: The expense of an operation that includes any money paid to an employee who performs work for the operation.

perceived value pricing: The process of adjusting a target menu price based on how management thinks a customer will perceive the price.

percentage: A number that expresses part of a whole in terms of hundredths.

perishable food: Food that has a short shelf life and is subject to spoilage and decay.

pie chart: A circle that is divided into pieces where each piece represents a percentage of the whole circle (or pie).

point-of-sale (POS) device: An electronic tool used to process customer orders, print guest checks, track customer and financial information, and generate financial reports.

portion control: The process of ensuring that a specific amount of food or beverage is served for a given price.

portion-controlled scoop: A volume measurement tool that consists of a handle with a fixed-size scoop at the end.

portion size: The amount of a food or beverage item that is served to an individual person.

pricing form: A tool often used to help calculate the AS cost of a menu item and establish a menu price.

product: The number that is the result of multiplication.

profit: The amount of money earned by an operation when revenue is greater than (>) expenses.

proper fraction: A fraction in which the numerator is smaller than the denominator.

purchase specification: A written form listing the specific characteristics of a product that is to be purchased from a supplier.

quotient: The number that is the result of division.

R

radius: The distance from the center of a circle to the edge of the circle and is equal to one half the diameter of the circle.

ratio: A mathematical way to represent the relationship between two or more numbers or quantities.

raw yield test: A procedure used to determine the yield percentage of a food item that is trimmed of waste prior to being used in a recipe.

reciprocal: A fraction that is the result of switching the places of the numerator and denominator in a fraction.

revenue: The total amount of money received by a foodservice operation from sales to customers.

rounding: The process of reducing the number of places in a decimal to achieve a certain degree of accuracy.

S

sales tax: A fee that must be collected from customers based on the requirements of state and local governments where the foodservice operation is located.

scaling: The process of calculating new amounts for each ingredient in a recipe when the total amount of food the recipe makes is changed.

scaling factor: The number that each ingredient amount in a recipe is multiplied by when the recipe yield is changed.

standardized recipe: A list of ingredients, ingredient amounts, and instructions for preparing a specific food item and amount of food.

standard profit and loss statement: A form that shows the revenue, expenses, and resulting gross and net profit (or loss) over a specific period of time.

statistics: A mathematical system for evaluating groups of numbers.

subtraction: The process of taking one number away from another number to find the difference.

sum: The number that is produced as the result of addition.

T

target food cost percentage: The percentage of food sales that a foodservice operation plans to spend on purchasing food.

target price: The price that a foodservice operation needs to charge for a menu item in order to meet its target food cost percentage.

U

unit cost: The cost of a product per unit of measure.

unit of measure: A fixed quantity that is widely accepted as a standard of measurement.

variable expense: An expense that increases or decreases based on the amount of sales.

volume: A measurement of the physical space a substance occupies.

weight: A measurement of the heaviness of a substance.

whole number: A number that is used for counting, such as 0, 1, 20, or 100.

yield: The total quantity of a food or beverage item that is made from a standardized recipe.

yield percentage: The edible-portion (EP) quantity of a food item divided by the as-purchased (AP) quantity and is expressed as a percentage.

Index

Page numbers in italic refer to figures.

yield, 95, *95*
  scaling recipes, 96–97, *97,* 234
yield percentage, 116–123
  calculations, 117–119, 236
  circle, *117*
  considerations, *121,* 121–123
  of fruits and vegetables, *247*
  on pricing forms, 157, *157, 249*
yield tests, 118–119, *119*

USING THE *CULINARY MATH PRINCIPLES AND APPLICATIONS* INTERACTIVE CD-ROM

Before removing the Interactive CD-ROM from the protective sleeve, please note that the book cannot be returned for refund or credit if the CD-ROM sleeve seal is broken.

System Requirements

To use this Windows®-compatible CD-ROM, your computer must meet the following minimum system requirements:
- Microsoft® Windows Vista™, Windows XP®, Windows 2000®, or Windows NT® operating system
- Intel® Pentium® III (or equivalent) processor
- 256 MB of available RAM
- 90 MB of available hard-disk space
- 800 × 600 monitor resolution
- CD-ROM drive
- Sound output capability and speakers
- Microsoft® Internet Explorer 5.5, Firefox® 1.0, or Netscape® 7.1 web browser and Internet connection required for Internet links

Opening Files

Insert the Interactive CD-ROM into the computer CD-ROM drive. Within a few seconds, the home screen will be displayed allowing access to all features of the CD-ROM. Information about the usage of the CD-ROM can be accessed by clicking on Using This Interactive CD-ROM. The Quick Quizzes®, Illustrated Glossary, Master Math™ Applications, Media Clips, Flash Cards, Forms and Tables, and ATPeResources.com can be accessed by clicking on the appropriate button on the home screen. Clicking on the American Tech web site button (www.go2atp.com) accesses information on related educational products. Unauthorized reproduction of the material on this CD-ROM is strictly prohibited.